Math Rules!

1st - 2nd grade
25 week enrichment challenge

by
Barbara VandeCreek

Graphic Design by Pam Jensen Pirk
©2001 Pieces of Learning
1990 Market Road
Marion IL 62959
polmarion@midamer.net
www.piecesoflearning.com
CLC0239
ISBN 1-880505-79-7
Printed in the U.S.A.

Acknowledgement

This editor especially thanks
Doug Gottes
at Pieces of Learning
for the many editing hours we spent agreeing and disagreeing in order to make *Math Rules!* a quality math resource for teachers of gifted students - assuring that the questions are not ambiguous and the answer key is correct.
We both have gold stars and scars
for completing 1st and 2nd grade!
We're looking forward to the 5th and 6th grade edition!

March, 2002
Thanks to Amber, Daniella, Joseph, and Mason
of Miss Lee's class in Ohio
for finding an error on page 61 in the 1st printing.

Kathy Balsamo
Editor

Teaching Suggestions for *Math Rules!*

Math Rules! is ready-made and easy-to-use. The problems have been teacher-and student-tested. Wording of problems has been carefully chosen, and answers have been verified. (But don't be surprised if some of your students put a twist on a problem that allows for a DIFFERENT answer!)

Reproduce and distribute the weekly **challenge problems worksheets** for individual students, partners, triads, small groups or for the whole class. Or place them in a Math Learning Center. For **1st Grade** there are **six challenge problems** for the week. They are marked with *numerals* for the grade level and *numbers* for the week.(**1: One** = First Grade, Week One). For **2nd Grade** there are **eight challenge problems** for the week. (**2: One** = Second Grade, Week One.)

Distribute or make available the **Challenge Problems Worksheets** at the beginning of the week. Expect their return by the end of the week. This gives time for thoughtful and reflective work and promotes student responsibility.

Have students identify how they *plan* to work (individually, with a partner, in a group) or how they *did* accomplish the work by circling the students at the top of each worksheet. If they worked together, all names can be placed on one worksheet.

Share problem-solving methods and answers with students (**drawing a picture, looking for a pattern, systematic guessing, role playing, using tables, working simpler similar problems, working backwards, using manipulatives, using technology**). It is important for students to recognize that problem-solving is like drawing - there may be more than one right way (process) to get the right answer. Individuals have a "style" preference. The process of uncovering an answer can be more significant than the correct answer. Post the commonly used strategies as headings on the bulletin board and cite problem examples under each heading. As weekly problems are solved, place examples of problems that use the posted strategies under the appropriate heading.

The *Stars* icon subjectively suggests the difficulty level of each problem - 1- 4. Each week it is possible to acquire 18 *Stars* "points" for First Grade and 24 *Stars* "points" for Second Grade. Teachers may choose to give only full credit for correct answers or partial scores (stars) for problems if the process is appropriate even though the computation is incorrect. Students may want to use individual charts to register their success. This encourages them to challenge themselves rather than compete with other students.

Enrichment, extra credit, or a regular assignment,
Math Rules! is **DIFFERENT**
and will energize your math curriculum and enthuse students.

Stars for *Math Rules!*

Math Rules! for Students

□ meets the wide range of math abilities and interests of 1st & 2nd grade **gifted math students**, in a gifted program, pull-out program, and in the regular classroom.

Many students reject math problem-solving because problems are too wordy. ***Math Rules!*** problems are not. However, solution-finding does demand discipline, flexibility, and creativity from students.

The Challenge Problems develop logic and reasoning skills and self confidence in the understanding of mathematics.

Math Rules! for Teachers

□ assures that students are regularly involved in significant critical thinking skills and problem-solving strategies.

Math Rules! is designed to meet the needs of teachers who are looking for an **organized, inclusive resource** that will provide a weekly significant consistent challenge throughout the school year.

25 weeks of challenging problems are offered for each grade level.

1: One through **1: Twenty-five**
and
2: One through **2: Twenty-five**

Math Rules! for the Math Curriculum

□ enriches a school's regular curriculum and students are **accountable** for their progress and time.

Students learn best by building upon what they already know. ***Math Rules!*** Challenge Problems are designed to help students use skills they learned in the regular curriculum and in the real world. Students solve most problems without new knowledge.

Goals of *Math Rules!*

Students will . . .

1. learn to value mathematics
2. become confident in their abilities
3. become mathematical problem solvers
4. learn to communicate mathematically
5. learn to reason mathematically

Essential Knowledge, Skills, Concepts and Standards in *Math Rules!* Challenge Problems

1. Number Sense, Operation, and Numeration
2. Computation - Addition, Subtraction, Multiplication, Division
3. Measurement
4. Estimation
5. Geometry
6. Fractions and Decimals
7. Quantitative Reasoning and Interpretation
8. Patterns, Relationships, and Algebraic Thinking
9. Spatial Reasoning
10. Statistics and Probability
11. Problem-Solving using Strategies and Tools
12. Understanding and Communicating Processes
13. Making Connections and Application to Everyday Experiences

Myself Partners Group

Math Rules!
1: One

Name ______________________

1. Make the next 2 shapes in the pattern.

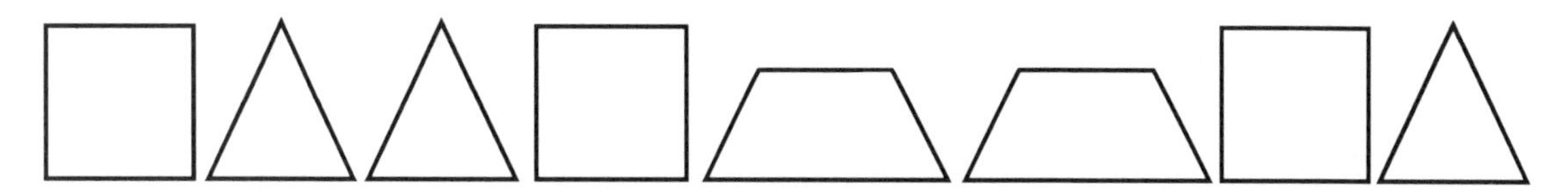

__________ __________

2. Put an X on the animals that have four legs.

3. This is one way to show 7.

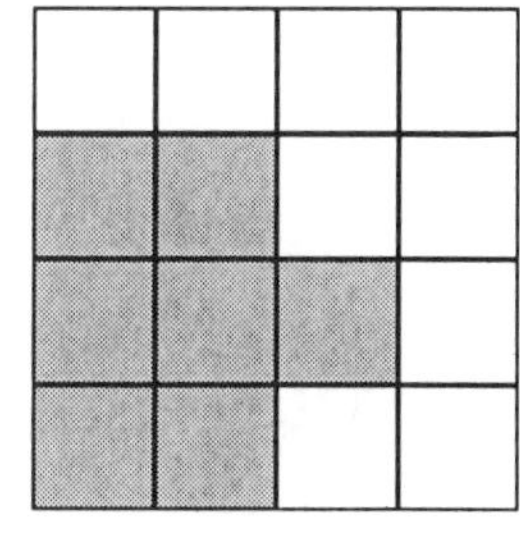

Fill in squares in other ways to show 7. The sides of your little squares must touch the side of another square.

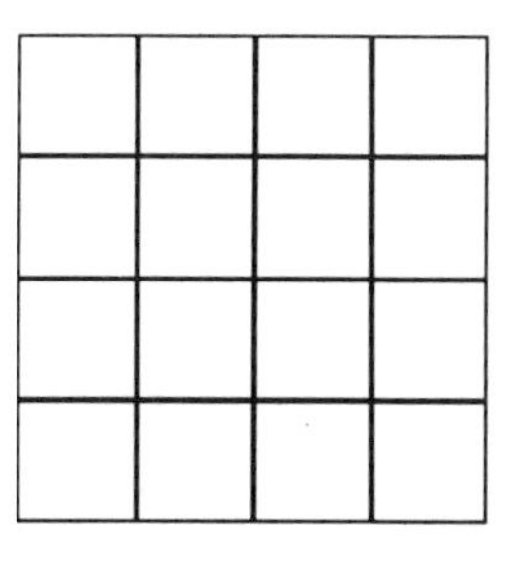

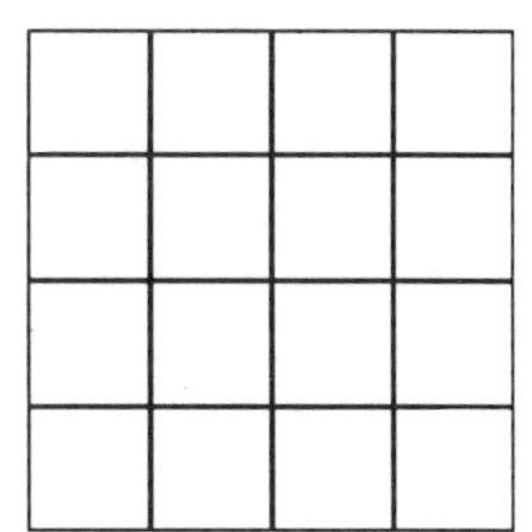
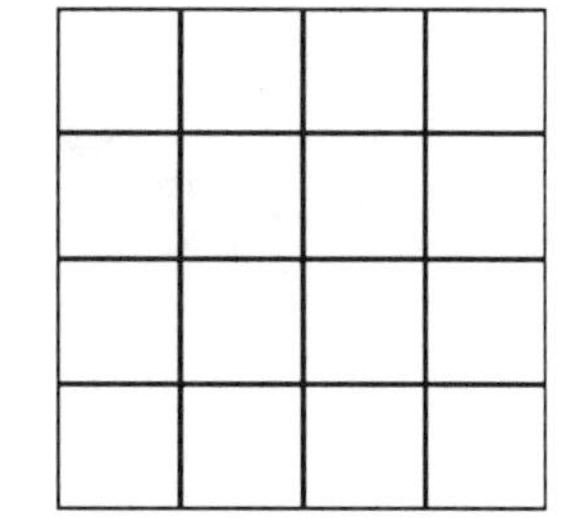
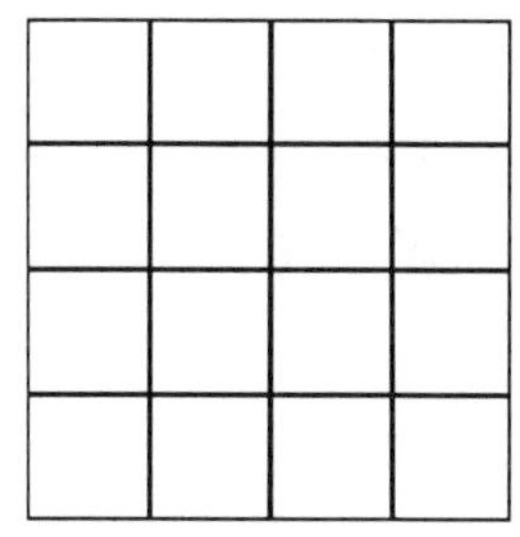

1: One

4. Will each bee have a flower?

YES NO

5. Mom gives Sue 10¢ a day for work she does at home each week.

How much money did she earn this week?

____

6. Follow the steps in this flowchart. Write your answer in the last box.

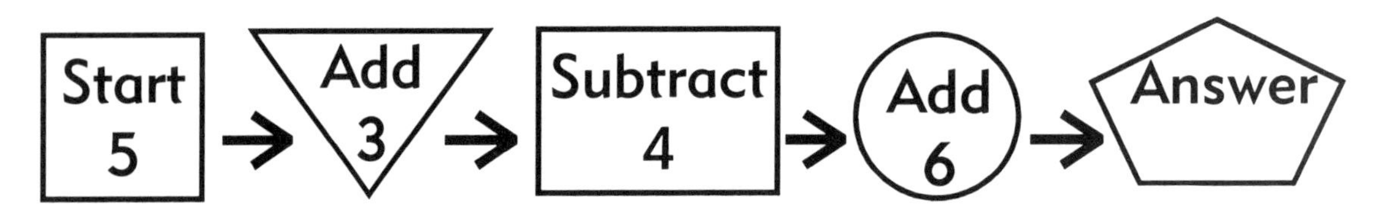

Myself Partners Group

Math Rules!
1: Two

Name ____________________

1. Mike wanted to play with number cards using 1 - 10. He had cards for 1, 6, 2, 8, 5, 7. Which cards are missing?

2. Read, then write the numbers.

two four six three nine

3. Four first graders had a race. Ben did not finish in first or last place. Josh came in third.

What place did Ben finish?

1 2 3 4

1: Two

4. Draw a line to match the letter and the number.

A B C D E F G H I J

first tenth second eighth sixth

5. Joe and Sarah took turns popping balloons.

 Joe popped these balloons.

Sarah popped these balloons.

If Sarah popped a balloon first, who popped the last balloon? Joe Sarah

6. Write a number sentence to show how far the frog jumped.

____

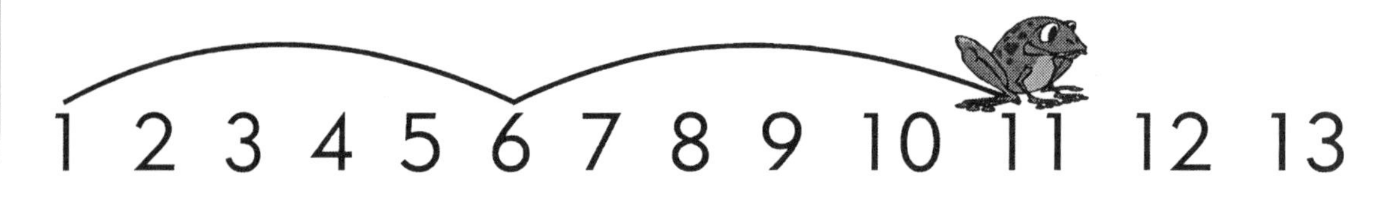

Myself Partners Group

Math Rules!
1: Three

Name ___________________________

1. Tara spent 24 cents on 3 stickers. Circle the stickers she bought.

12¢ 11¢ 8¢ 4¢ 3¢

2. Circle the bucket that is heavier.

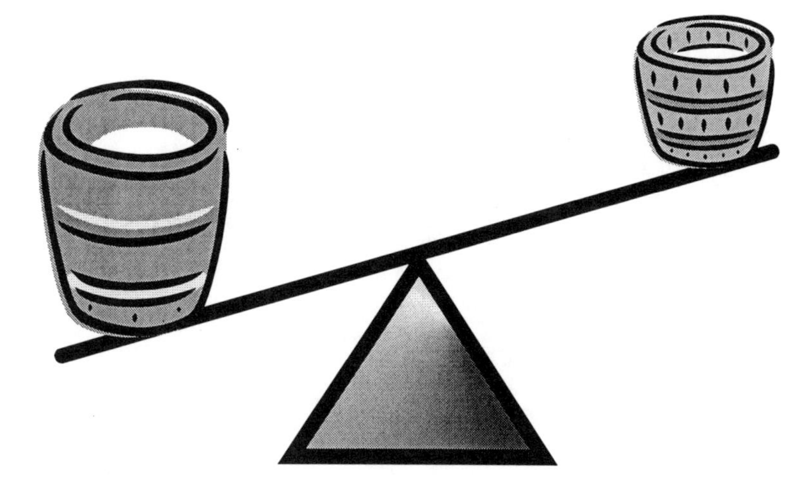

3. Find the pattern. Then fill in the empty triangle to fit the pattern.

1: **Three**

4. A coin is under one nut shell. It is not under the second shell, but it is to the right of the second shell. Circle the nut shell that has a coin under it.

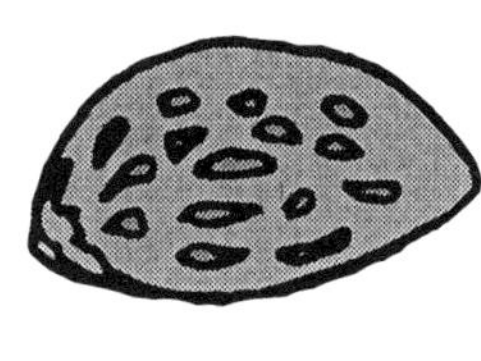

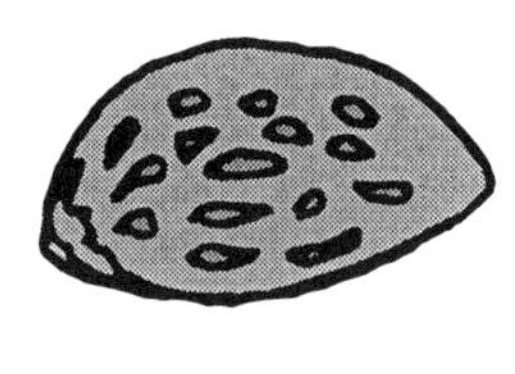

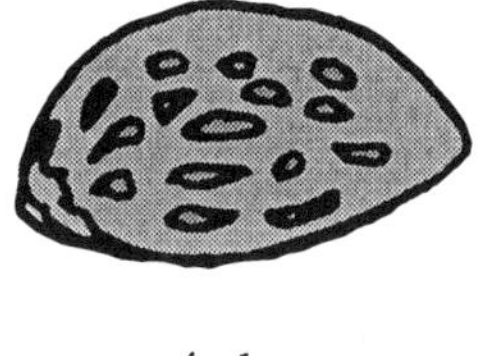

1st 2nd 3rd 4th

5. How many stars are on the space shuttle?

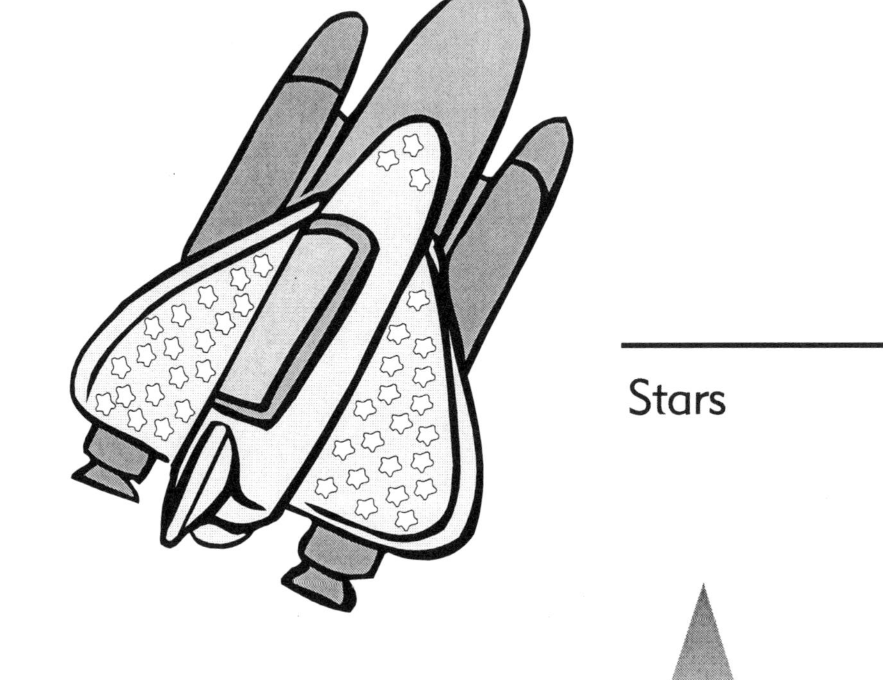

__________ Stars

6. Find the missing number on the star so that the sum of the numbers equals 23. __________

Myself Partners Group

Math Rules!
1: Four

Name ____________________

1. Ask 12 children in your class which color they like best. Use tally marks to show how many.

RED	BLUE	YELLOW	GREEN

How many children like each color?

RED BLUE

YELLOW GREEN

2. How many squares are in the design?

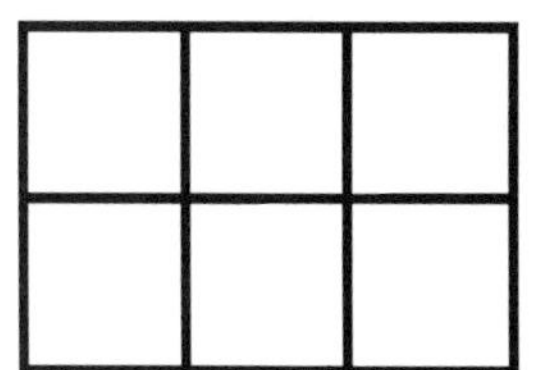

__________ squares

3. How many dollar bills do you need to pay for the umbrella if you do not have any coins?

_____ Dollars

1: Four

4. Put the numbers 1, 2, 3, 4, 5, 6 in the triangles so that the sum of each side is 10. Use each number only once.

5. Steve has a handful of jelly beans to share with 2 friends. X out some jelly beans from his hand and draw them in Ryan's and Eric's hands so all the boys have the same.

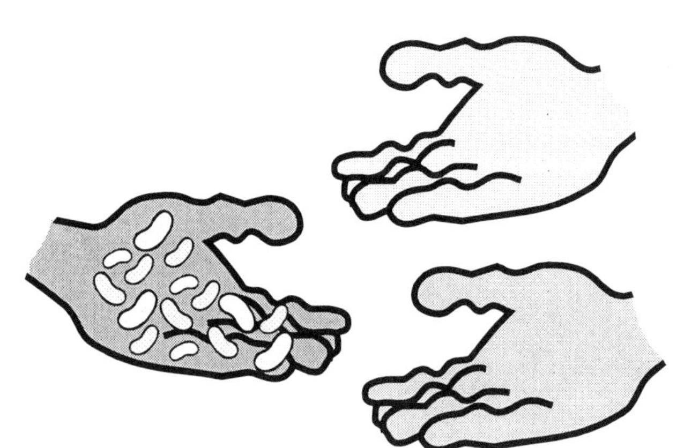

6. Look carefully at the designs and count.

Now write the numbers and add.

Example:

★ + ✧ = ?

3 + 2 = ❖ (5)

❖ + ✧ = ? ________

⊙ + □ = ? ________

□ + ★ = ? ________

✧ + ✧ = ? ________

Math Rules!
1: Five

Name ______________________

1. Circle the two boys that are just the same.

2. How much money do these coins equal?

Answer __________

3. Find the score for each team.

Lions	0	1	2	1	0	2	1	0	3
Tigers	1	0	0	4	0	1	0	2	1

Final score - Lions _______
Final score - Tigers _______

1: Five

4. Count the tally marks.
Then connect the dots in order.

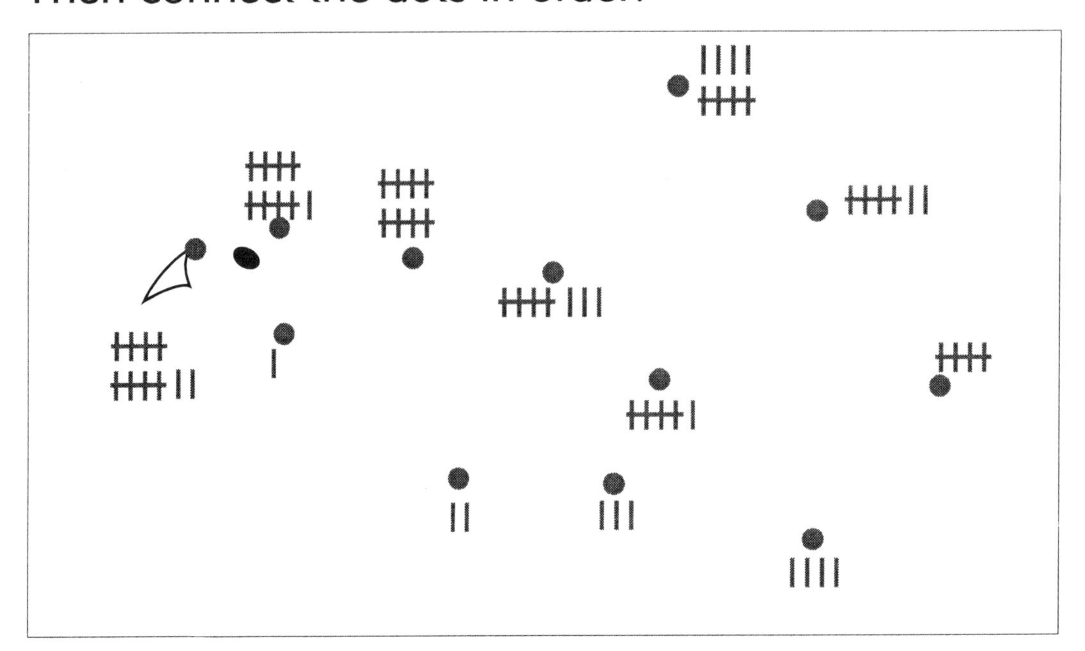

5. Draw the design that comes next.

6. Add these numbers to find the <u>sum.</u> To make your job easier, look for numbers that add to ten.

	8	9	5	8
3	7	5	6	3
5	7	4	2	5
7	1	4	2	7
+5	+2	+1	+5	+5

Myself Partners Group

Math Rules!
1: Six

Name ____________________

1. About how many paper clips will fill the space?

a. 5 b. 8 c. 12

2. Divide this circle exactly in half.

(Idea: Hold the paper up to the light and fold it in half. Draw a line.)

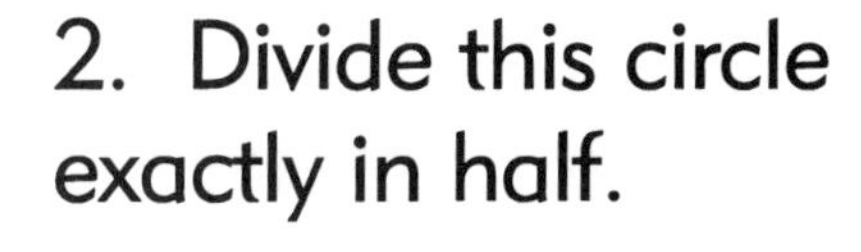

3. Use a calculator. Press keys to find the answer.

ON 2   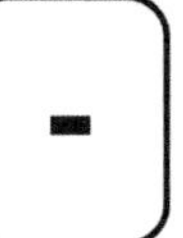 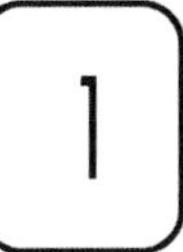 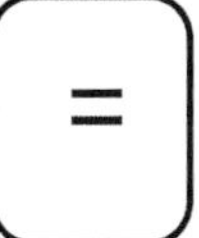 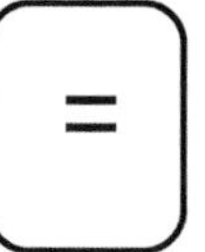 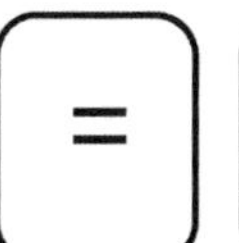 

The number on the display is 

1: Six

4.

How many ❂ ______ How many ❄ ______

How many ❃ ______ How many ✳ ______

5. What is the temperature on each thermometer?

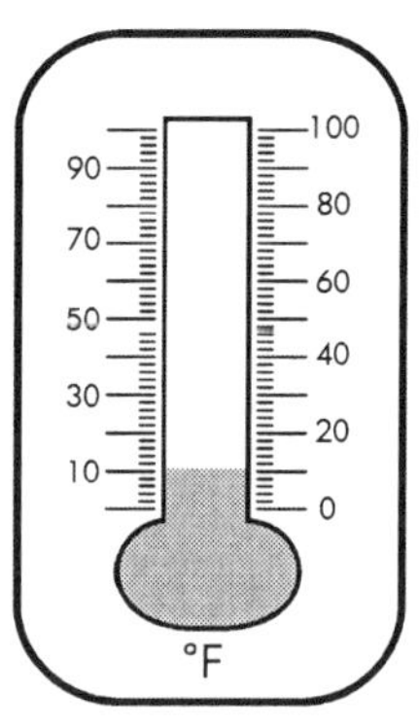

______° F ______° F

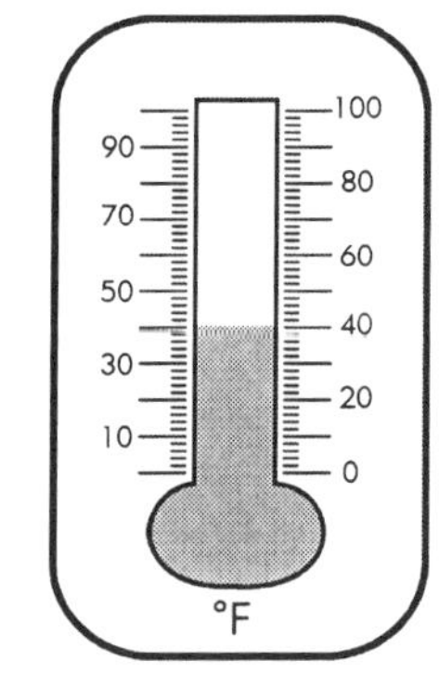

6. On both thermometers, color the middle part red to show the temperature.

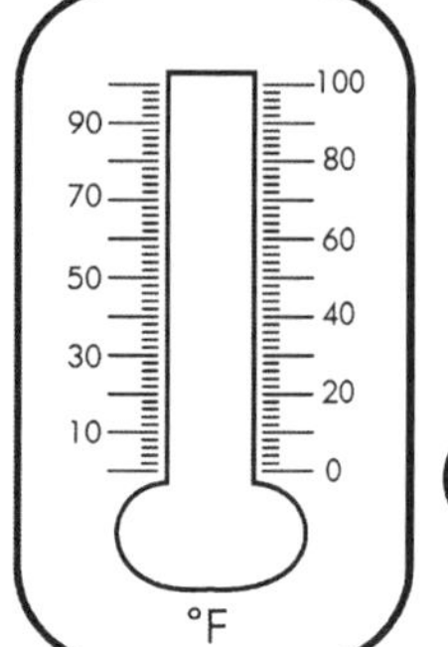

0° F

30° F

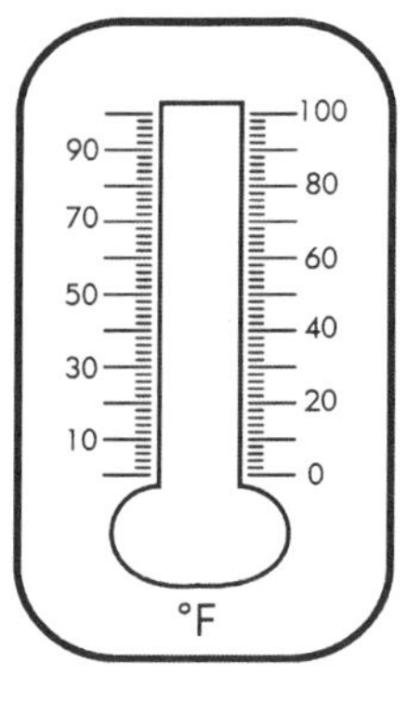

Math Rules!
1: Seven

Name ______________________

1. Fill in the circles with the numbers 1, 2, 3, 5, 6, 7. All three numbers in a line must equal 12. Use each number just once.

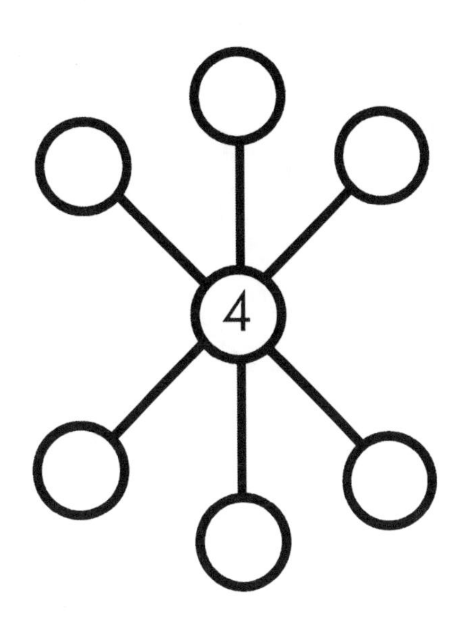

2. Write the numeral for
   5 tens + 3 ones + 2 hundreds ______________

3. Tell what you will find when you go . . .

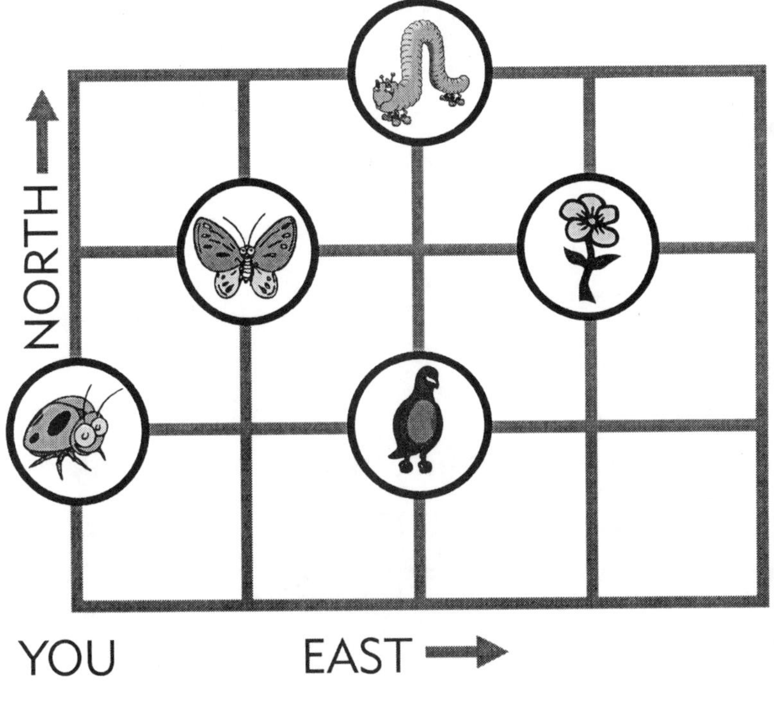

1 block EAST,
2 blocks NORTH

____

3 blocks EAST,
2 blocks NORTH

____

0 blocks EAST,
1 block NORTH

____

1: Seven

4. Make true number sentences.

Circle

GREATER or LESS

	IS		THAN
10	GREATER	LESS	8
76	GREATER	LESS	67
27	GREATER	LESS	30
500	GREATER	LESS	350
99	GREATER	LESS	100
1	GREATER	LESS	0
143	GREATER	LESS	163

5. Nick starts school

at  and lunch time is at 

How much time passes before Nick has lunch? __________

6. Jane fed three squirrels in her yard at home. By the time her mom called her to come inside, there were eight squirrels. How many squirrels joined the feeding? __________ Squirrels

Math Rules!
1: Eight

Name ______________________

1. Find the code that will give the phone numbers for the special stores.

a. BIKES GO

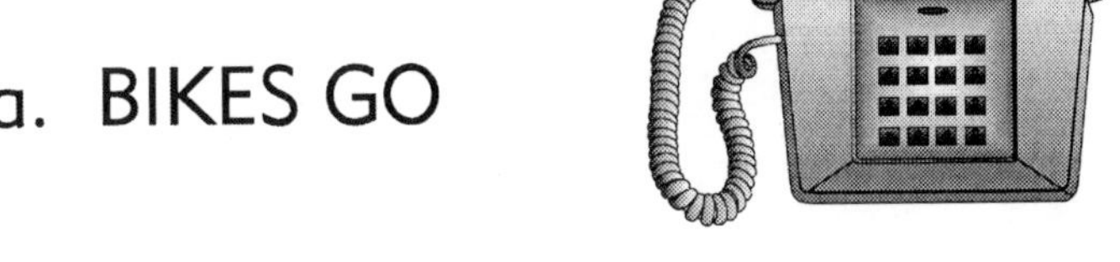

___ ___ ___ - ___ ___ ___ ___

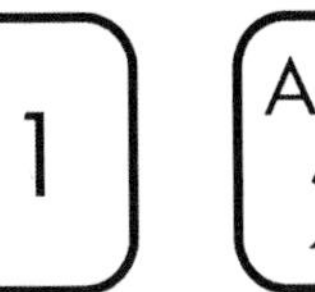

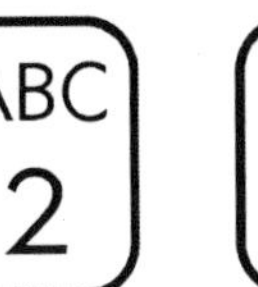

b. OH SWEET

___ ___ ___ - ___ ___ ___ ___

c. U LV MATH

___ ___ ___ - ___ ___ ___ ___

2. Find the number that goes in the first square.

$\square - \boxed{5} + \boxed{9} + \boxed{3} = \boxed{14}$

3. How many ● are

in the triangle? ________

in the square? ________

in the hexagon? ________

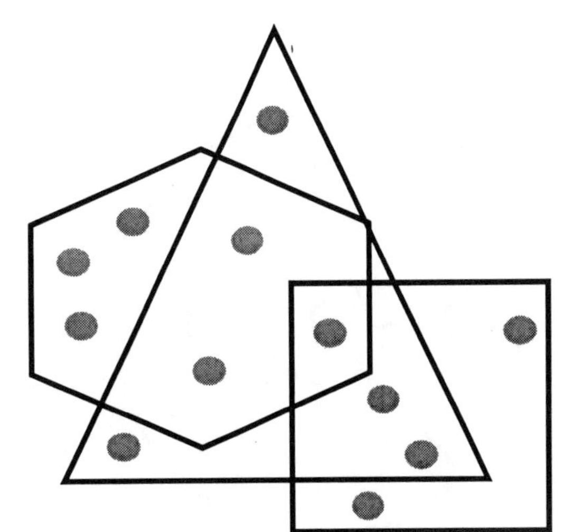

1: Eight

4. Circle the numbers whose digits add up to 24.

9348 5144 4848 685

6783 9025 8372 6666

3429 7557 1928

5. About how tall is the boy?

________ cm

About how long is the boy's arm?

________ cm

About how long is his shoe?

________ cm

6. Write the missing numbers in the subtraction circle.

Math Rules!
1: Nine

Name ______________________

1. Boxes of beads: Show different ways to put 5 beads in 2 boxes.

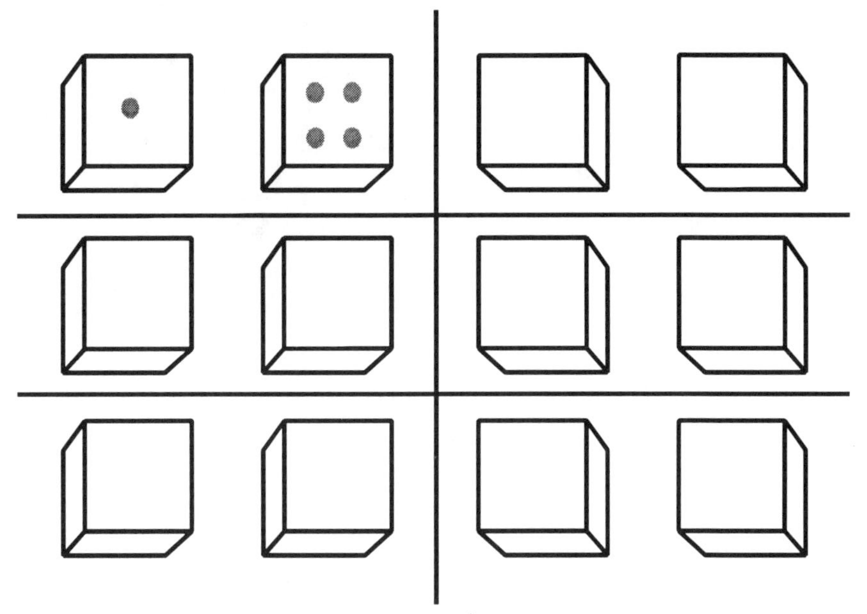

2. Which weighs more: a [crayon] or a [eraser] ?

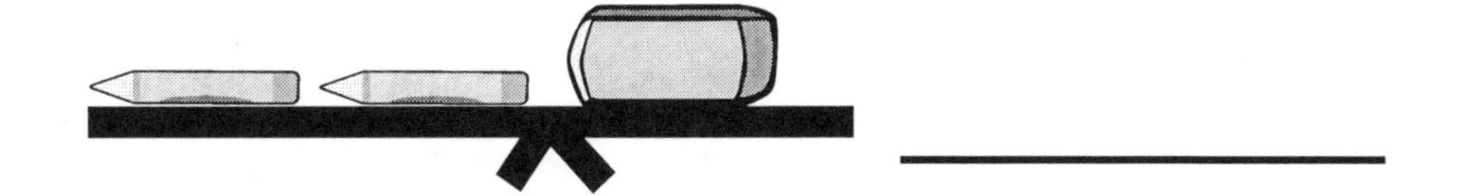

____

3. About how long is the tree branch?
Circle your answer.

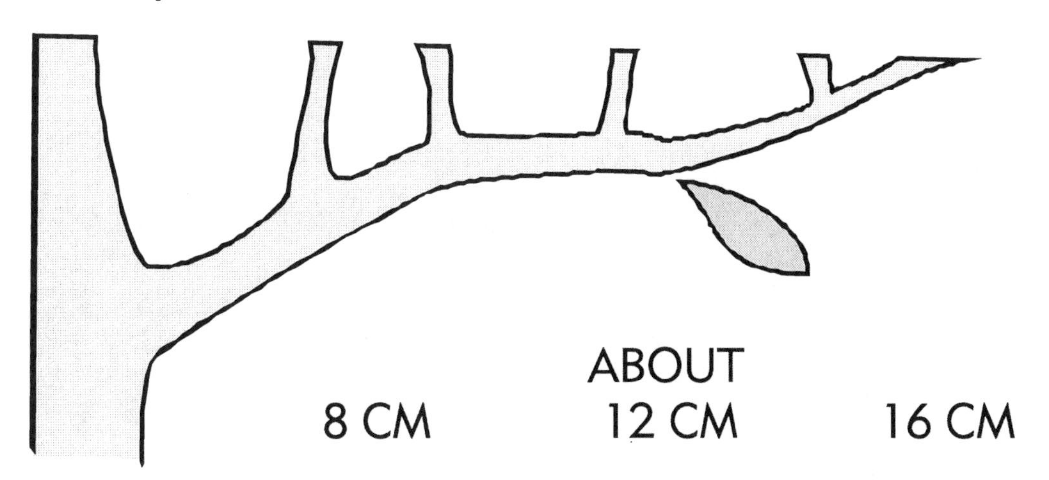

ABOUT
8 CM 12 CM 16 CM

1: Nine

4. Use the clues to find where Miles, Aaron, Grandma, Bryan, and Ann live.

- ★ **M**iles does not have to use the stairs or the elevator.
- ★ **G**randma lives 2 floors above Miles.
- ★ **A**nn stopped to say "Hi," to Grandma. Then she took the elevator 2 more floors.
- ★ **Aa**ron lives between Miles' grandma and Miles.
- ★ When **B**ryan goes to play with Ann he walks up one set of stairs.

Write the initials of the people in the windows.

5. Which is the shortest road the bus can take to school? ______

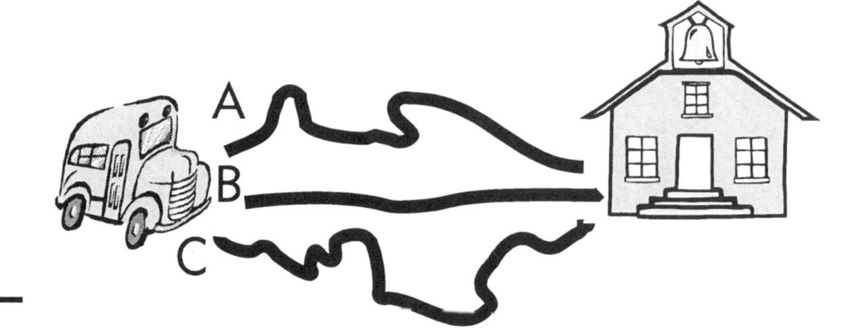

6. Martin liked to put his baseball cards in order on the table. With each new row he put one less card than the row before it. Martin started with 10 cards in the first row. Then he put down 9 cards in the second row. How many cards did he line up in the fifth row? __________ How many cards did Martin put down in the first 5 rows? __________

Math Rules!
1: Ten

Name ______________________

1. Make a bar graph showing the kinds of pets children in the first grade class owned.

Jack	cat	Emily	dog
Andy	dog	Marcus	bird & cat
Christy	dog	Jeriel	dog
Shelly	cat	Kim	turtle
Jake	cat	Tasha	bird
Amber	dog	Jenna	2 cats
Cara	turtle	Cindy	cat
Kim	cat	Steve	dog
Mike	bird	Jana	turtle

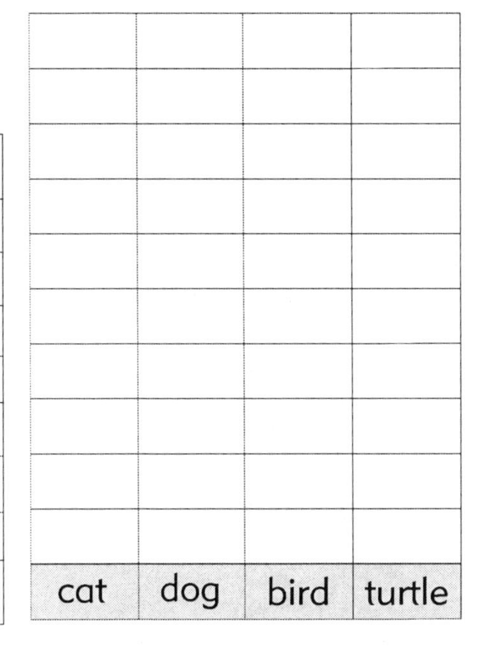

2. The number line has the answers to fill in the table.

↑↓

	Start on	Move	Equals
a.	3	-2	1
b.	-3	+2	
c.	6	-4	
d.	-3	+6	
e.	-6	+4	
f.	1	-5	
g.	-3	+6	

6
5
4
3
2
1
0
-1
-2
-3
-4
-5
-6

3. Start at the right side of the row. Circle the seventh ✸ and second ✸.

1: Ten

4. An angle is made where two straight lines meet. Circle the angles you find on this boat.

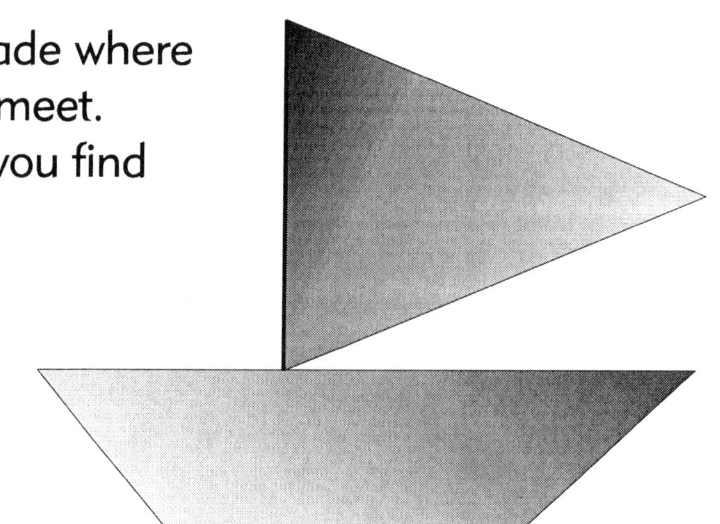

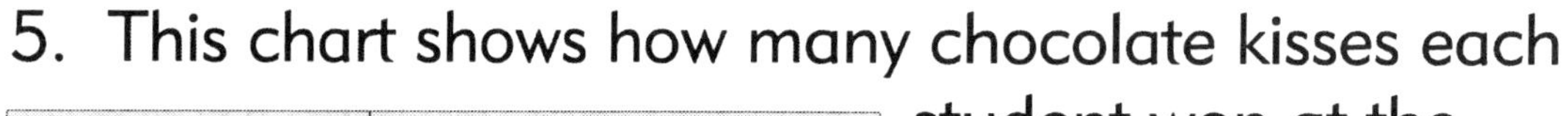

5. This chart shows how many chocolate kisses each student won at the school carnival. Circle which student won more than Roma but less than Lee.

Student	Kisses Won
Roma	4
Ted	6
Holland	3
Lee	8

Roma Ted

Holland Lee

6. Write the numbers in the boxes where they belong.

25 63 28 4
18 49 32 21
9 99 38 2

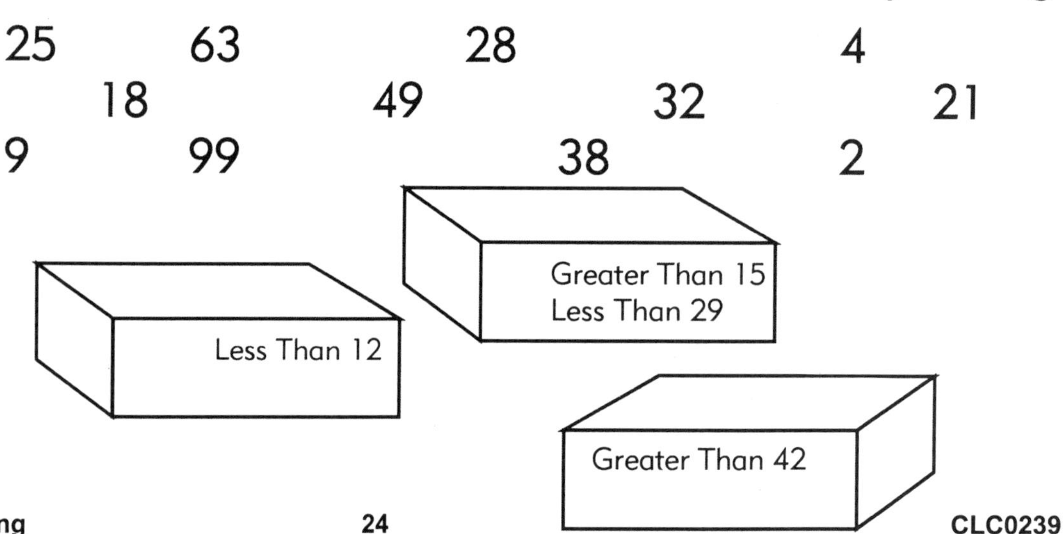

Math Rules!
1: Eleven

Name ______________________________

1. Write the hidden math word.

SUBTRACT ______________________________

ZERO ______________________________

FIFTH ______________________________

SUM ______________________________

TIME ______________________________

2. Put 2, 3, 4, 6, 7, and 8 in the circles so that the sum of each line is 19. Use each number only once.

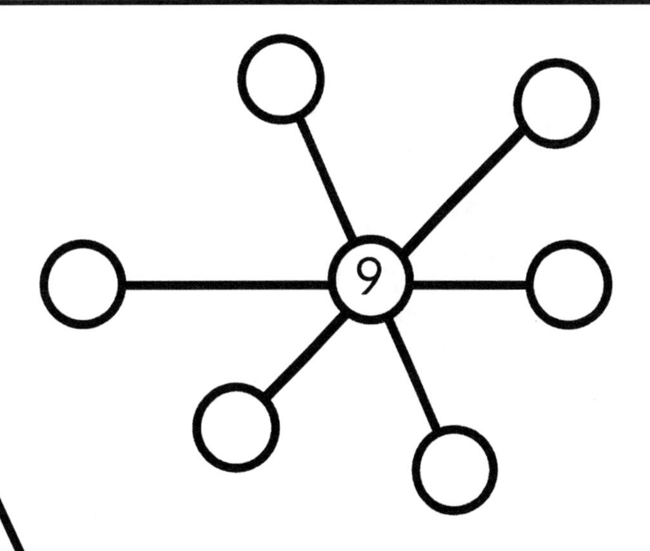

3. How many triangles can be found in this star?

______ Triangles

1: Eleven

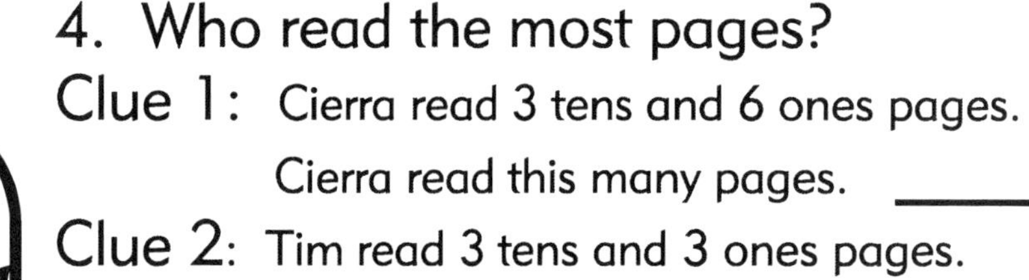

4. Who read the most pages?

Clue 1: Cierra read 3 tens and 6 ones pages.

Cierra read this many pages. ________

Clue 2: Tim read 3 tens and 3 ones pages.

Tim read this many pages. ________

Clue 3: Ted read one page more than Tim read.

Ted read this many pages. ________

Clue 4: Kelly read 5 pages fewer than Cierra.

Kelly read this many pages. ________

Circle the student's name who read the most pages.

Tim Ted Kelly Cierra

5. Color the shapes that can be folded so both sides match.

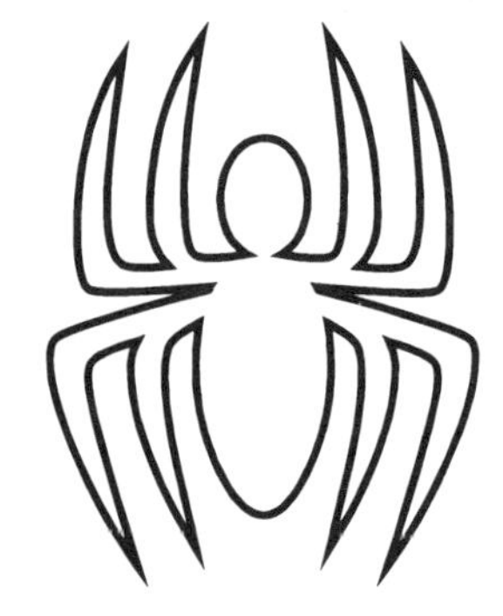

6. How many squares did it take to make this shape?

____

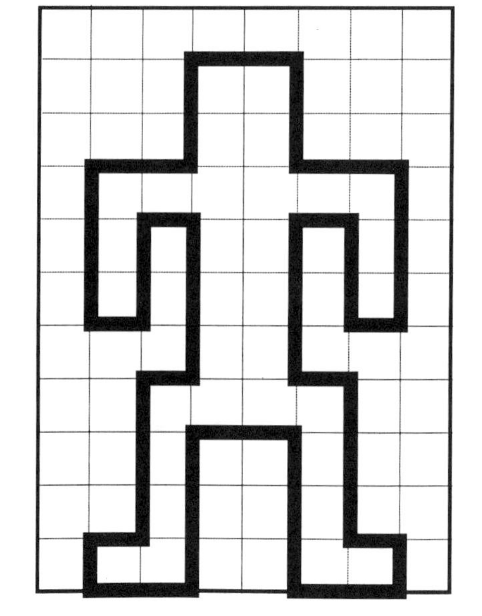

Math Rules!
1: Twelve

Name ____________________

1. Four children were throwing frisbees to see whose frisbee could land the farthest.

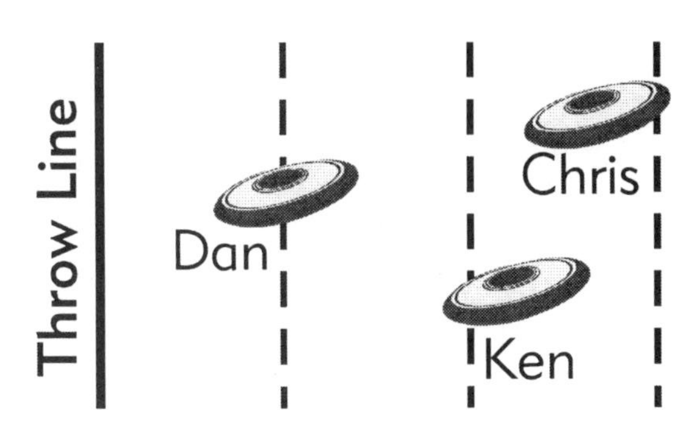

Dan's landed 2 meters from the throw line.

About how far did Alan's land from the throw line?

3 meters   8 meters   12 meters

2. Count by tens but begin with 4. Cross out the wrong numbers and write the correct number above it.

4 14 26 34 44 58 68 74 81 94

3. If your sister baked 22 cookies and you ate 2 cookies, how many cookies did you eat?

_______ Cookies

1: Twelve

4. Count backward by 2s. Cross out wrong numbers and write the correct numbers above them.

32 31 28 27 24 22 21 18

15 14 12 10 7 6 4 2

5. RECYCLED TOYS

You have $1.00 to spend. What can you buy? You can circle more than one toy. Spend the entire dollar.

6. Use a calculator to find the answer to this problem.

9509 - 1774 = __________

Turn the calculator upside down. The word you see is the opposite of "buy."

What is the word? ____________

Myself

Partners

Group

Math Rules!
1: Thirteen

Name ______________________________

1. An angle is a *corner* where two straight lines meet.

HOW MANY ANGLES ARE THERE IN . . .

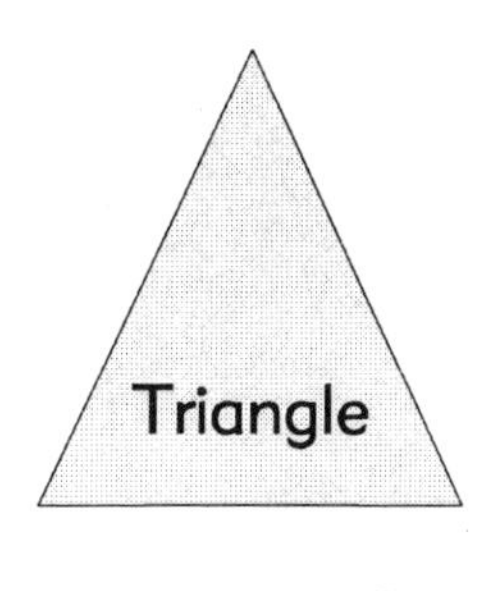

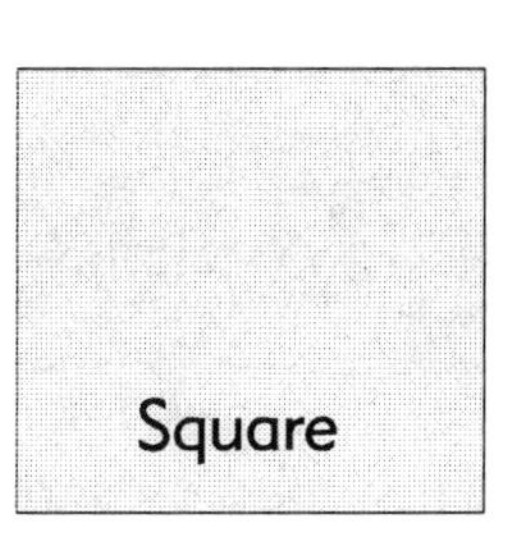

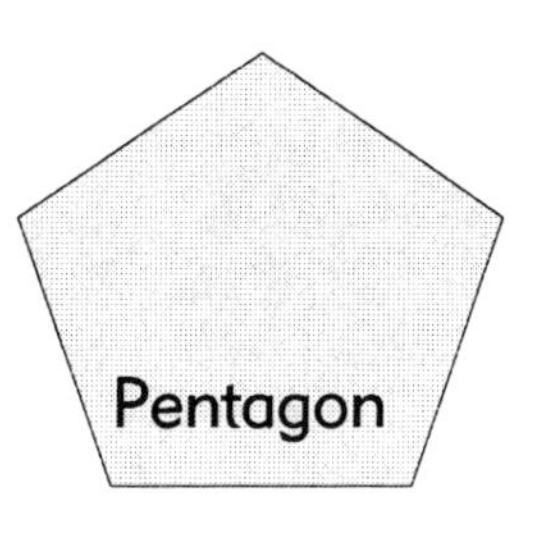

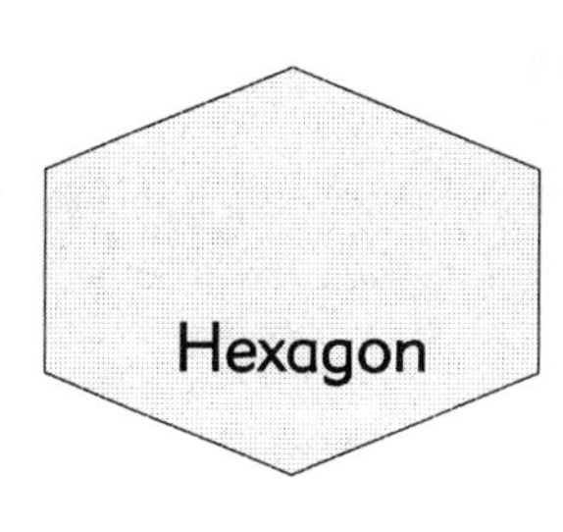

_____ ∠s   _____ ∠s   _____ ∠s   _____ ∠s

2. A coin has two sides. The side with a face is called "heads," and the other side is called "tails." Flip a coin in the air 10 times, and let it land on the floor. Mark on the chart each head and each tail you get.

HEADS										
TAILS										

WHICH SIDE LANDED UP MOST OFTEN?

Heads   Tails

3. Circle the word that makes this sentence true.

The length of a classroom is measured in . . .

grams   liters   meters   millimeters

4. Out of the four listed fruits how many total pieces of fruit are in this list? 

6 apples
8 pears
5 blueberries
9 carrots
4 oranges

5. Finish drawing the pencil so that it is 13 centimeters long from the end of the eraser to the lead tip.

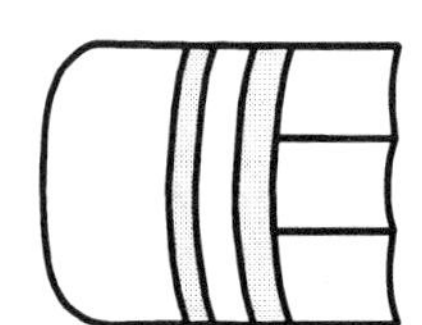

6. How many 2-digit numbers can you make using the digits 3, 5, and 8? Use each digit only once in a number. I can make _____ numbers.

Myself Partners Group

Math Rules!
1: Fourteen

Name ____________________

1. Where is the package of M & M's®? Circle it.

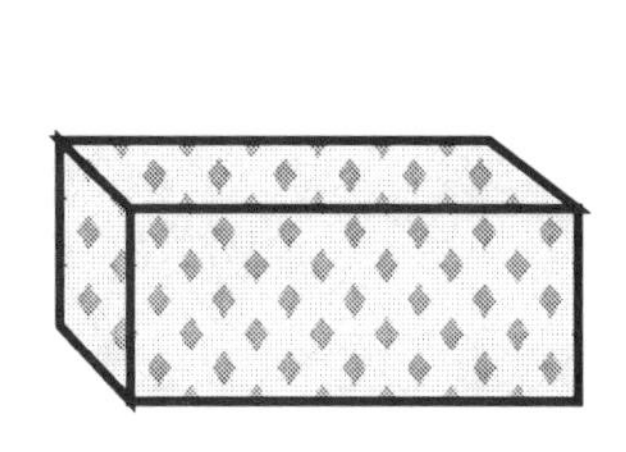

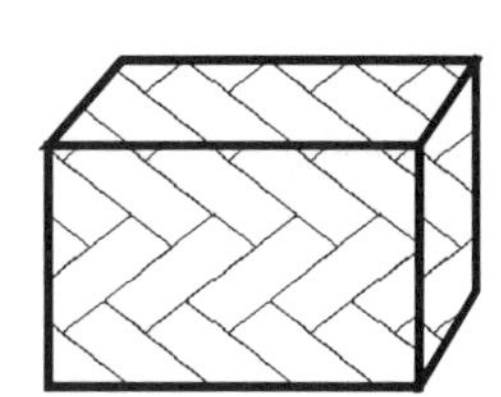

Clues:

1. TWIX® are in the box on the right.
2. SNICKERS® are in the box next to the PEANUT BUTTER CUPS®.
3. M&M's® are in the box between the PEANUT BUTTER CUPS® and the TWIX®.
4. The SNICKERS® are in the box on the left.

2. How many 0's and what number comes next?

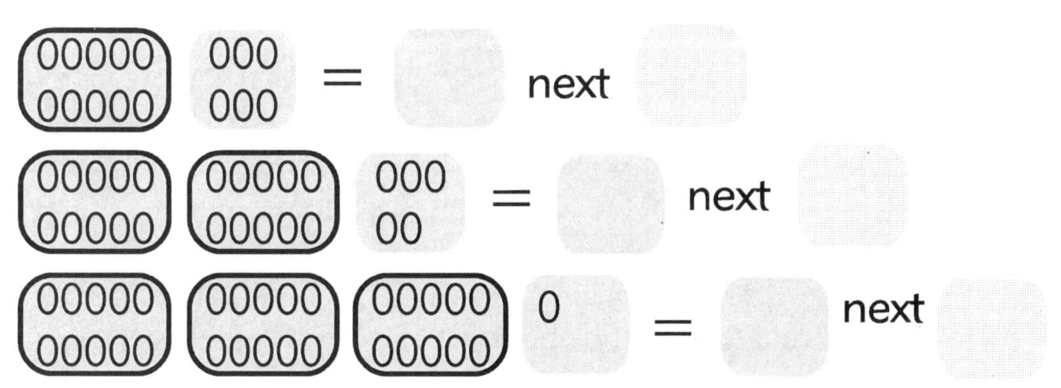

3. Skip count by 5's.

5 10 15 ______ ______ ______ ______

1: Fourteen

4. How many pennies are here? ________

How many pennies are in each row?

____

How many pennies are in each column?

____

5.

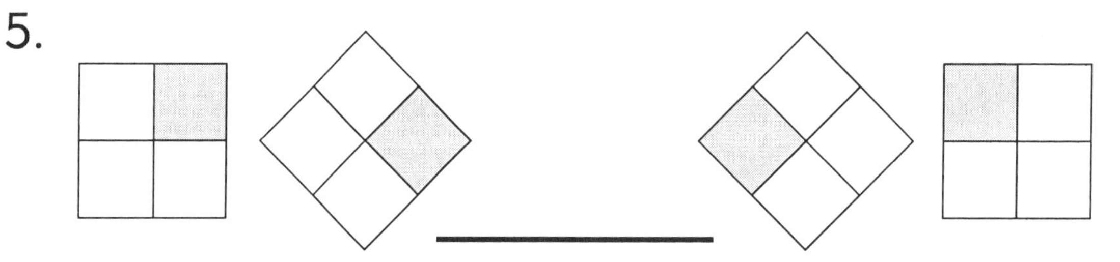

Which of these 3 squares belongs in the empty space above?

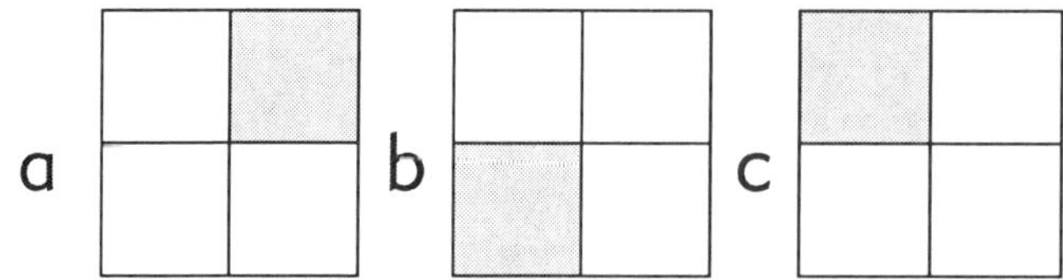

6. How many apples could you buy with 20 pennies?

____

How many pennies are left over?

____

Math Rules!
1: Fifteen

Name ____________________

1. How many times can you subtract this number 9 from this number 36?

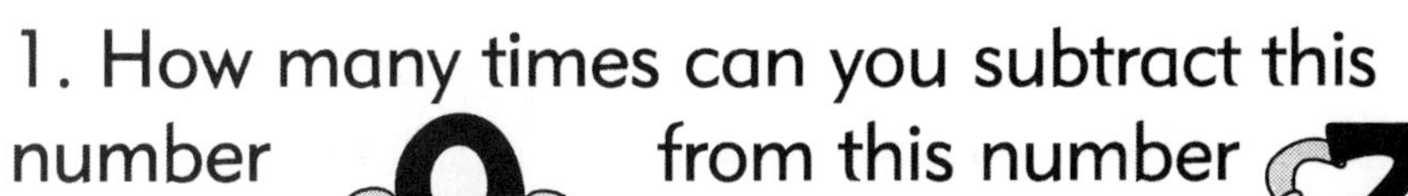

_____ times

2. Fit the four pieces into the puzzle. When the puzzle is done, the numbers must add up to 15 in all directions.

9
3 5 7
1

Odd numbers of things can not be put in pairs.

Even numbers of things can be put in pairs.

3. If all odd numbers are dotted and all even numbers are striped, will an odd number plus an even number be dotted or stripped? ______________

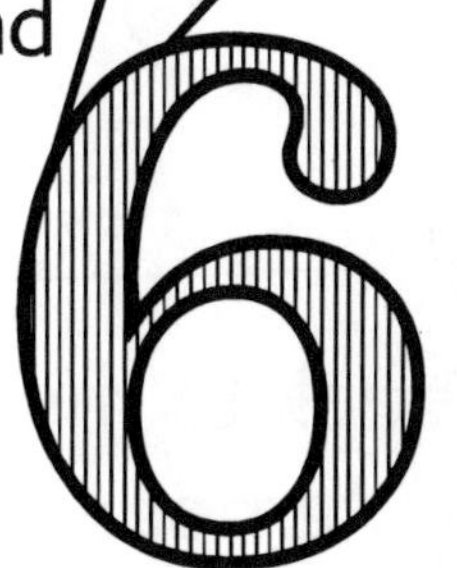

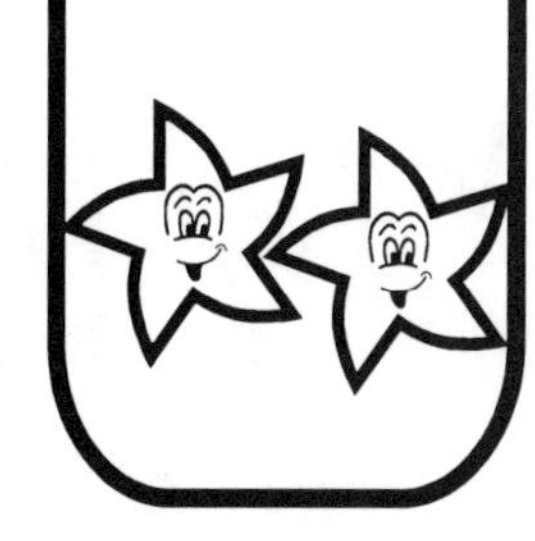

1: Fifteen

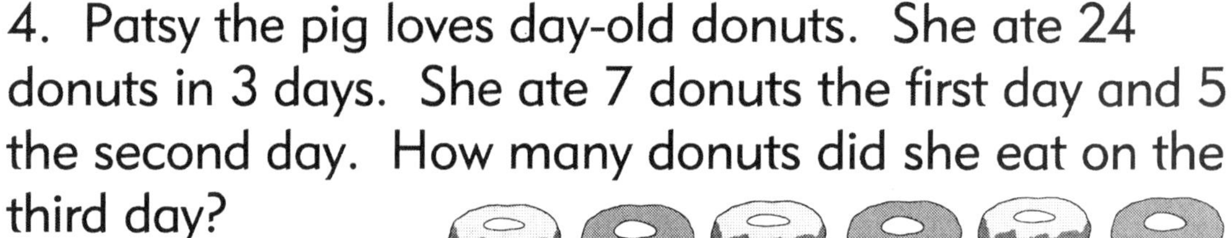

4. Patsy the pig loves day-old donuts. She ate 24 donuts in 3 days. She ate 7 donuts the first day and 5 the second day. How many donuts did she eat on the third day?

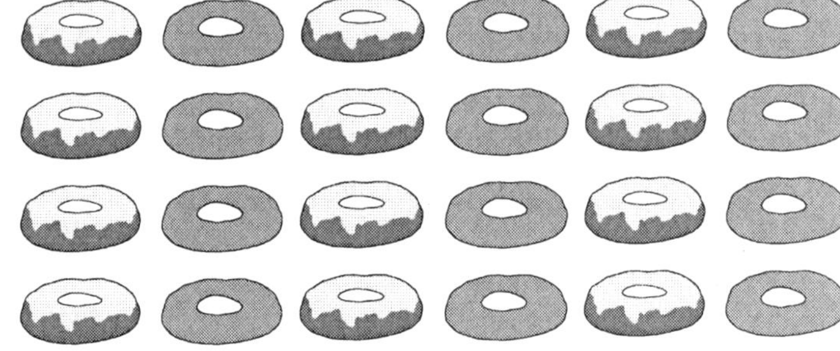

________ donuts on the 3rd day

5. Look at the thermometers. Write the temperature on the thermometer on the line next to it.

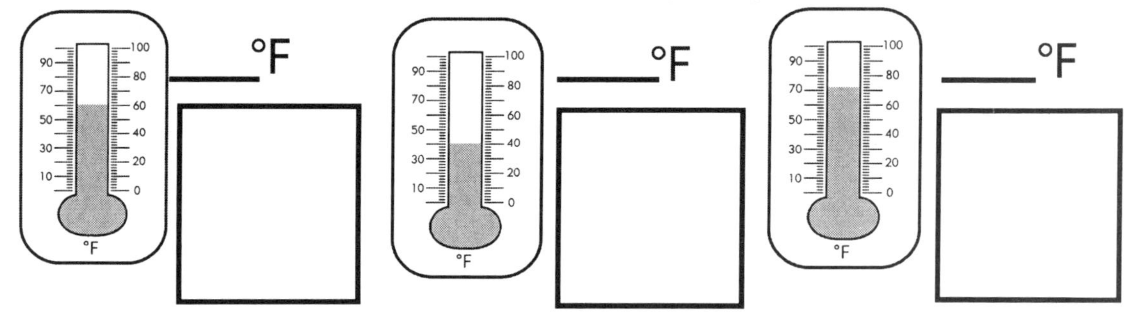

____°F ____°F ____°F

6 If the temperature is 70° or above, it is a warm day.

If the temperature is 50° to 70°, it is a cool day.

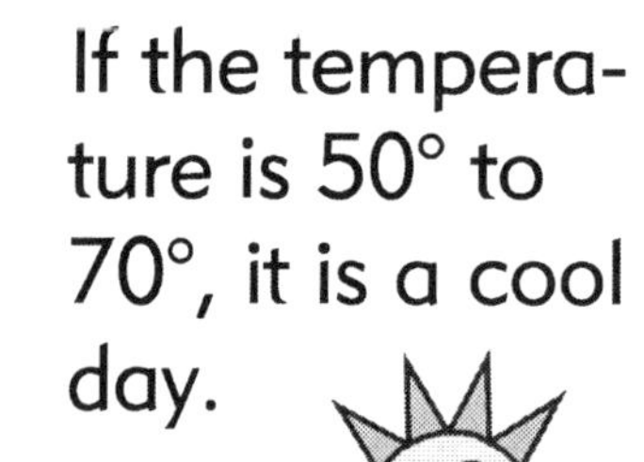

If the temperature is 50° or below, it is a cold day.

Go back to the thermometers in problem 5 and draw these temperature symbols in the correct boxes.

Myself Partners Group

Math Rules!
1: Sixteen

Name ______________________

1. Draw enough ● to show the heavier side of the scale.

2. How far is it around each shape?

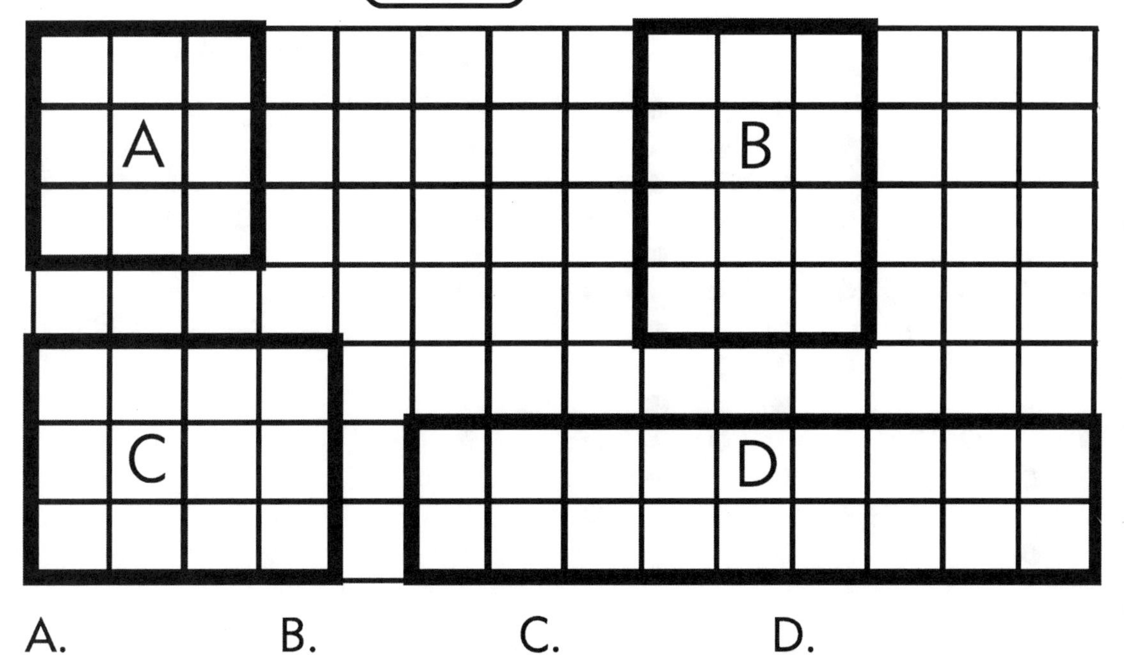

A. ________ B. ________ C. ________ D. ________

3. Show the time on each clock.

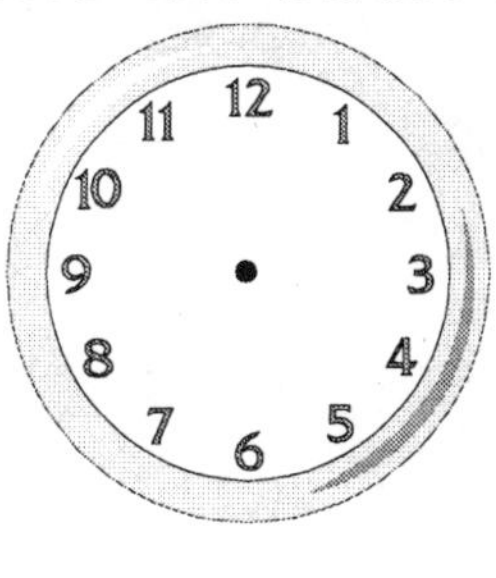

8 o'clock

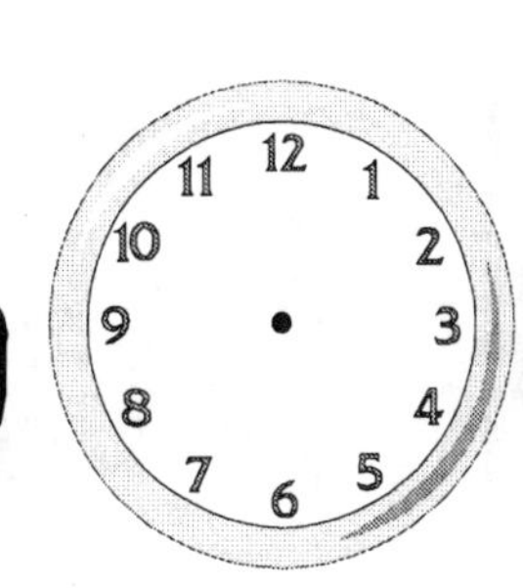

3:30 o'clock

1: Sixteen

4. How many spaces are there between the spokes of two 10-spoked bike wheels? ________ spaces.

5. Circle groups of ten dots. Write how many tens and how many ones are left.

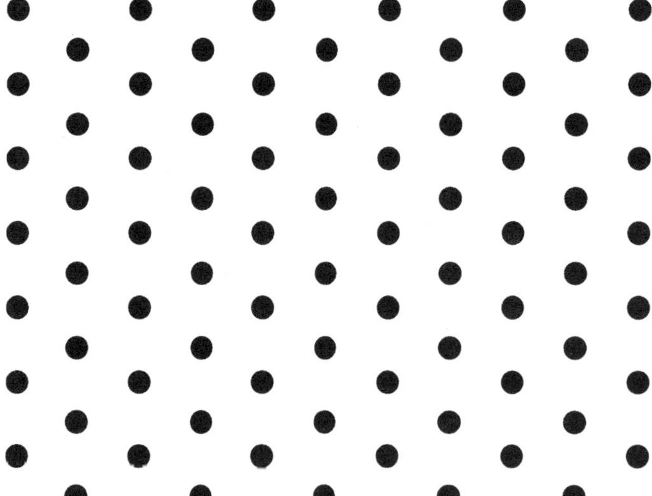

______ 10's

______ ones

6. Show how much money the sun glasses cost. Mark an X over the coins you do not need.

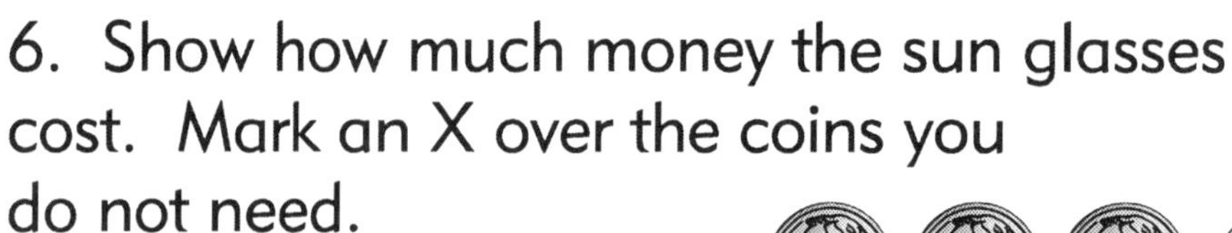

Myself Partners Group

Math Rules!
1: Seventeen

Name ______________________

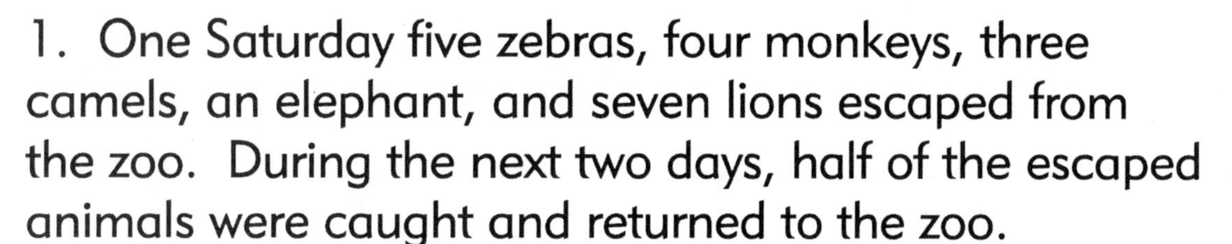

1. One Saturday five zebras, four monkeys, three camels, an elephant, and seven lions escaped from the zoo. During the next two days, half of the escaped animals were caught and returned to the zoo.

How many animals were still free on Tuesday? ______

2. Figure out how many 3 ounce cookies there are in a dozen.

______________ cookies

3. Circle the star(s) that appear to be the same size and same shape as the first star.

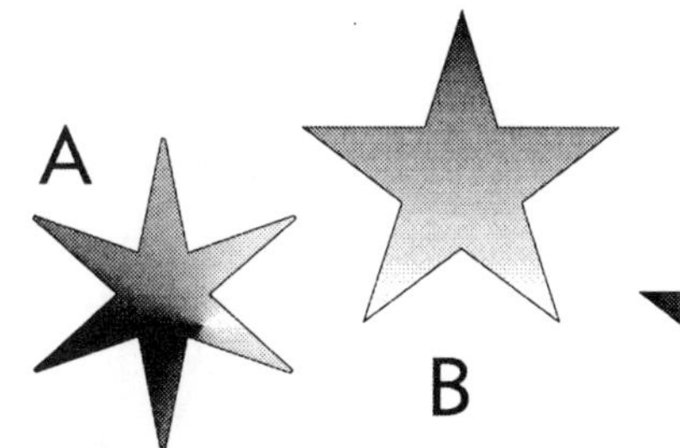

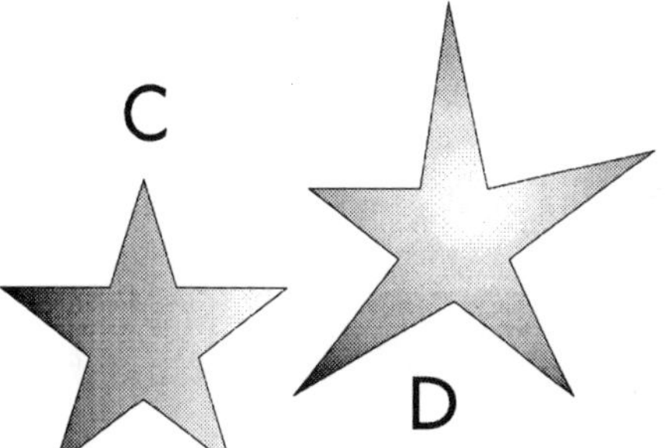

1: Seventeen

4. About how many dimes will fit in the purse?

About __________ dimes.

5. The square below is covering up a number. Write the correct number in the square.

$20 + \square + 15 = 43$

6. How many dots are inside, outside, and on the line?

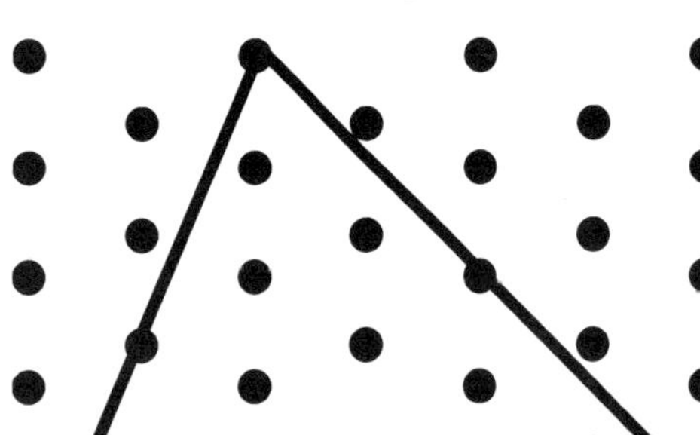

_______inside

_______outside

_______on

inside _______

outside _______

on _______

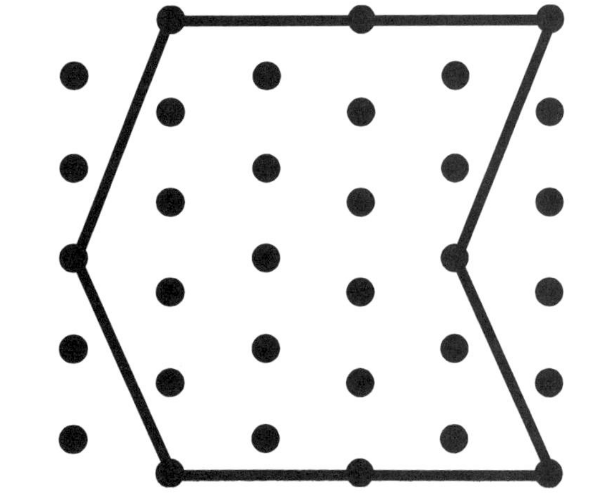

Math Rules!
1: Eighteen

Name ____________________

1. One DECIMETER is the same as ten centimeters.

Measure each side of the shapes below. Add the the lengths together. Is the length around the shape more than or less than 2 decimeters?

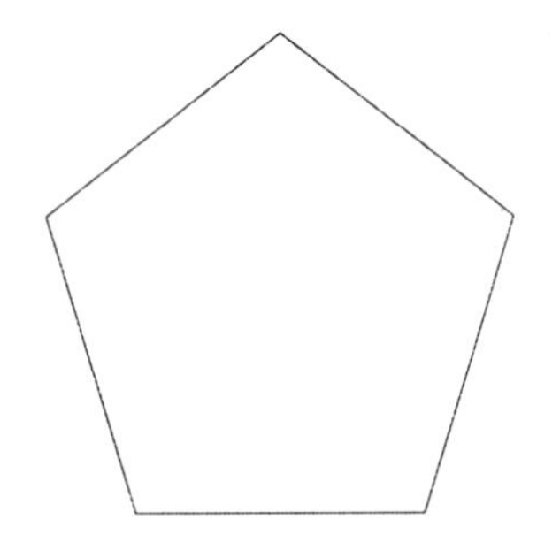

More than 2 decimeters
Less than 2 decimeters

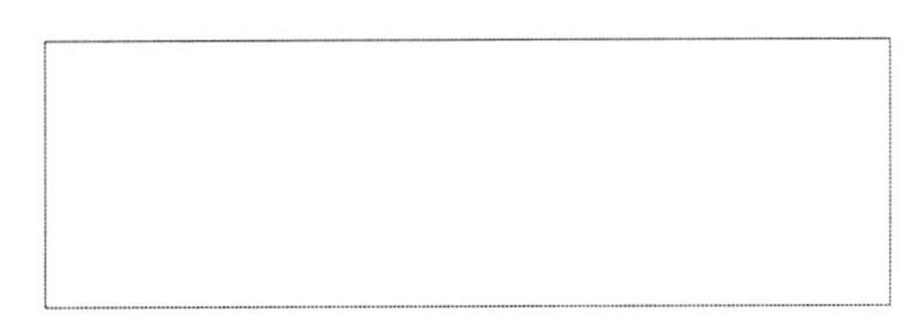

More than 2 decimeters
Less than 2 decimeters

2. How much money is here all together? ______ ¢

3. Add.

9	9	4	3
4	9	5	8
+5	+2	+6	+7

1: Eighteen

4. How many days are in 2 weeks?

____

FEBRUARY 2001						
S	M	T	W	T	F	S
				1	2	3
4	5	6	7	8	9	10
11	12	13	14	15	16	17
18	19	20	21	22	23	24
25	26	27	28			

5. Circle the day of the week VALENTINE'S DAY is on in February 2001.

Sunday Monday Tuesday Wednesday

Thursday Friday Saturday

6. Here is a map of Joe's city. How many different paths are there for Joe to go to school without going on the same road twice?

Myself Partners Group

Math Rules!
1: Nineteen

Name ______________________

1. Use the tax table to find how much each person spent.

STUDENT	COST		TAX		TOTAL
Jeff	53¢	=	______	=	______
Zoe	85¢	=	______	=	______
John	10¢	=	______	=	______
Greg	25¢	=	______	=	______
Ashley	79¢	=	______	=	______

Each Sale	Tax
.16 to .17	.01
.18 to .34	.02
.34 to .50	.03
.51 to .67	.04
.68 to .83	.05
.84 to 1.00	.06

2. The number in the *ones* place tells if a number is odd or even. Color ODD numbers blue. Color EVEN numbers red. Write any EVEN or ODD number to complete the pattern and color them.

34 28 51 44 96 35 62 80 17 14 58 59 6

3. Julie threw 5 darts and scored 16 points. All darts landed on a number on the board. On what numbers could her darts have landed?

____ ____ ____ ____ ____

1: Nineteen

4. Put a "¢" or "$" next to the numbers so that the sentence makes good sense.

New bluejeans cost 18 ____.

Mom gave Ted 75 ____ for school lunch.

A computer store sold the game CD for 24 ____.

5.

38 39 40 41 42 43 44 45 46 47 48 49 50 51 52 53 54 55 56 57 58

a. Is 43 closer to 40 or 50? ________

b. Is 58 closer to 50 or 60? ________

c. Is 54 closer to 50 or 60? ________

d. Is 39 closer to 30 or 40? ________

6. A  eats 5  every day.

How many flies does the spider eat in a week?

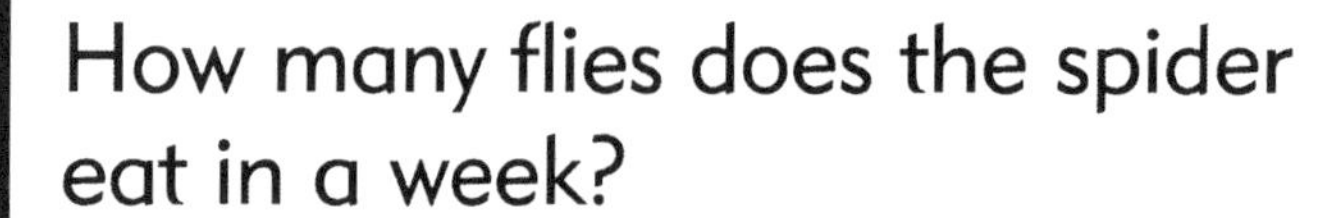

S M T W Th F S

____

Math Rules!
1: Twenty

Name ______________________

1. A very long time ago the people of ancient Babylon used different symbols for numbers than the ones we use. Look at their numbers.

V	1	VVVV	4	VVVV VVV	7	<	10
VV	2	VVV VV	5	VVVV VVVV	8	<V	11
VVV	3	VVV VVV	6	VVVVV VVVV	9	<VV	12

Write our number for VVVV
VVVV ____________

Write our number for <<V ____________

Write our number for <VVV
VV ____________

Write our number for <<<VV ____________

2. Write our number for the number word.

seventy-two ______
fifty-three ______
forty-four ______
ninety-five ______

3. This is the size of a card in a regular deck. About how many whole cards will it take to cover most of this math paper?
(do not overlap.)

1: Twenty

4. The first graders gave a play for parents. It started at 10:30 and was 2 hours long. Draw hands on the clock to show when the play ended.

5. The graph shows the kinds of bikes in the neighborhood.

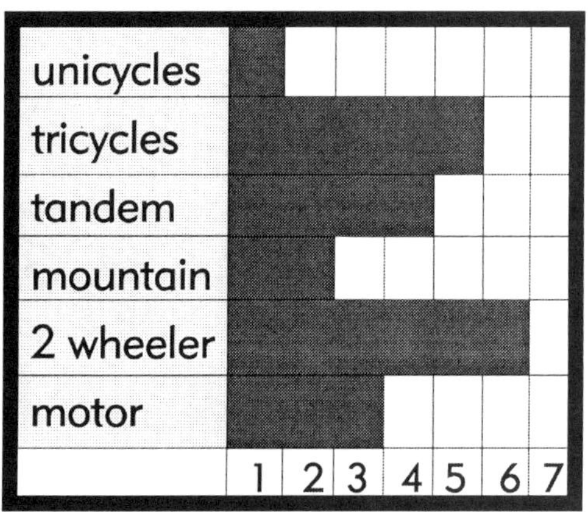

Write the number:

a. How many 2 wheelers?

b. Which kind of bike are there 3 of ? ____________

c. How many total bikes have 3 wheels? ____________

____

6. Write the letters on the CORRECT cards below.

Fourth R Fifth I Second E Sixth F

First T Seventh I Third R Eighth C

Myself Partners Group

Math Rules!
1: Twenty-one

Name ______________________

1. Find the path to match the ball with sports equip-ment.

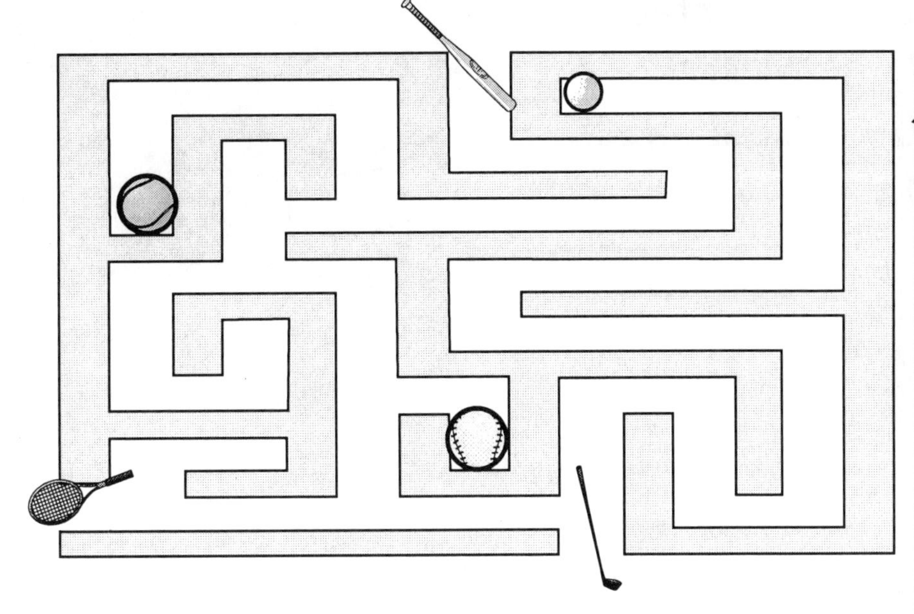

2. Joey, Kyle, Jim, and David have some money. Figure out how much they have in all.

- ★Joey has a dime.
- ★Kyle has a quarter.
- ★Jim has 5¢ more than Kyle
- ★David has as much as Kyle.

Together the four boys have __________ .

3. Here are 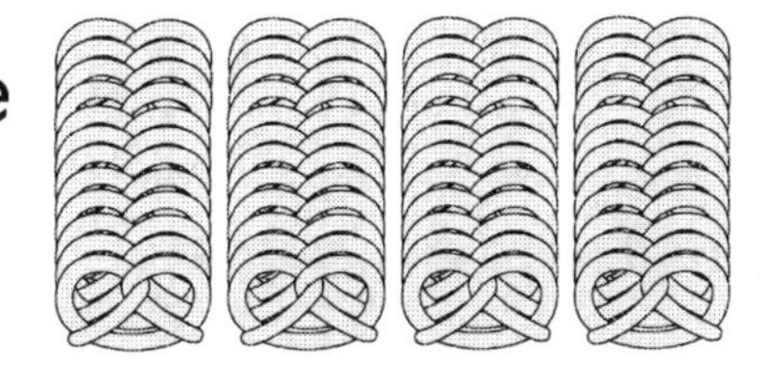 48 pretzels

About how many pretzels are here? 

10 pretzels 20 pretzels 30 pretzels

1: Twenty-one

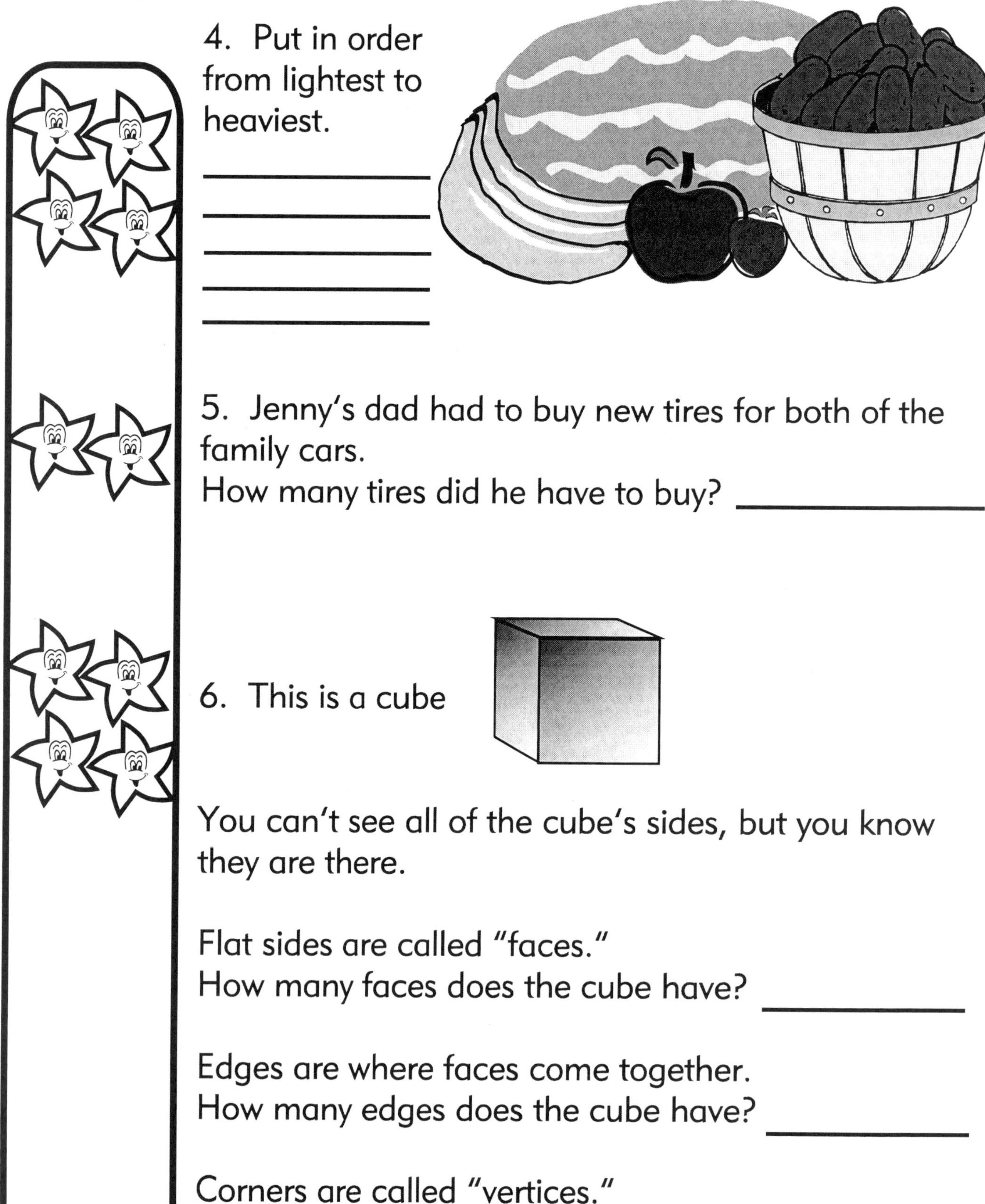

4. Put in order from lightest to heaviest.

____

____

____

____

____

5. Jenny's dad had to buy new tires for both of the family cars.
How many tires did he have to buy? ______________

6. This is a cube

You can't see all of the cube's sides, but you know they are there.

Flat sides are called "faces."
How many faces does the cube have? ______________

Edges are where faces come together.
How many edges does the cube have? ______________

Corners are called "vertices."
How many vertices does a cube have? ______________

Math Rules!
1: Twenty-two

Name ______________________

1. This scale shows the pumpkin weighs ____________ pounds.

2. This is a tricky maze. Look carefully. Both paths end on same number.

Count by 2s. Color this path yellow.

Count by 3s. Color this path blue.

2	4	8	33	62	41	27	29
3	6	9	12	19	13	15	11
51	8	10	15	37	24	22	20
13	15	12	18	21	24	17	35
5	24	14	16	18	27	40	38
45	57	24	36	20	30	33	39
30	28	26	24	22	48	36	56
32	34	36	38	40	42	39	0

On what number do both paths end? ______

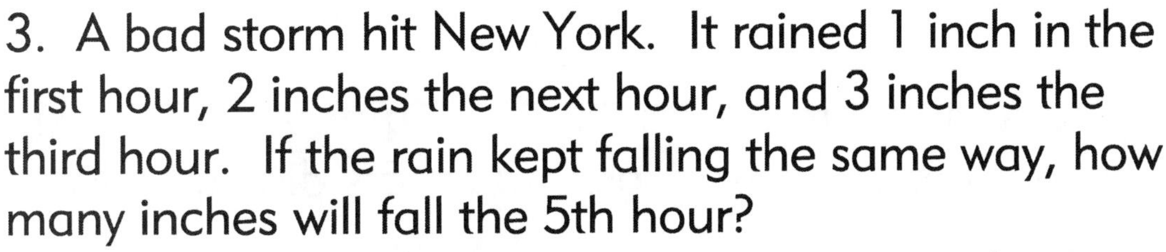

3. A bad storm hit New York. It rained 1 inch in the first hour, 2 inches the next hour, and 3 inches the third hour. If the rain kept falling the same way, how many inches will fall the 5th hour?

____ inches

1: Twenty-two

4. When we find the nearest 10 of a number, that number is called "rounded."

23, ROUNDED TO THE NEAREST TEN, IS 20.

27, ROUNDED TO THE NEAREST TEN, IS 30.

Round the numbers to the nearest 10.

33 ________ 38 ________
56 ________ 52 ________
69 ________ 64 ________
24 ________ 81 ________

5. Match to show what you would use to measure.

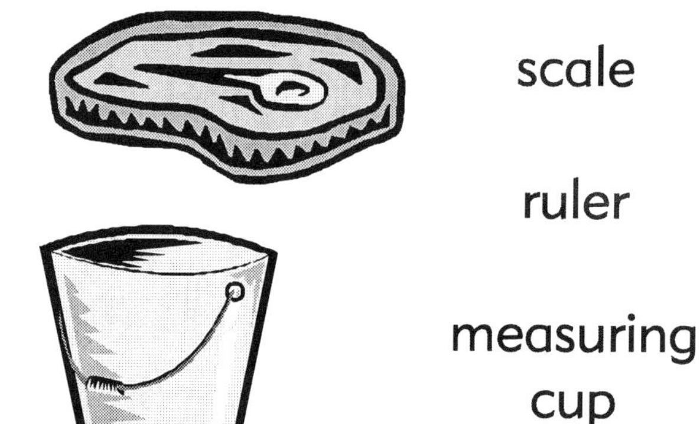

scale

ruler

measuring cup

6. Jackie had three pet cats. During one year all her cats had three kittens. How many cats did she have at the end of the year? ______ cats.

Math Rules!
1: Twenty-three

Myself Partners Group

Name ______________________________

1. One of the cheerleaders in this row is holding her pom-poms in the wrong pattern. Who is it? X it.

2. Use the money in the purse to buy what is for sale. Circle the coins you will get as change. Write it in the blank.

Popcorn 65¢
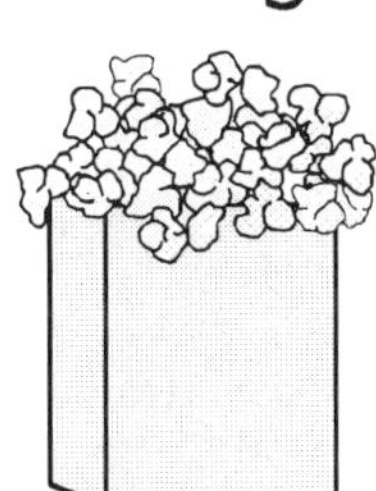

________ ¢

Yo-Yo 20¢ and Whistle 20¢

________ ¢

Paper 90¢ and Pencil 5¢

________ ¢

3. The rabbit must get back across the river. Help the rabbit count as it steps on the rocks.

1: Twenty-three

4. This bicycle is about 5 feet long. Circle the sentence that makes sense.

The bike is about 5 feet high.
The bike is about 1 foot high.
The bike is about 3 feet high.

5. Put > (greater than), < (less than), or = (equal to) in the square to make the number sentence true.

8 + 7 ☐ 20 - 5

6. Look at the number in the calculator's display window.

- How many tens? ________
- How many ones? ________
- How many hundreds? _____

Math Rules!
1: Twenty-four

Name ____________________

1. Subtract across and subtract down. Fill the squares with the missing numbers.

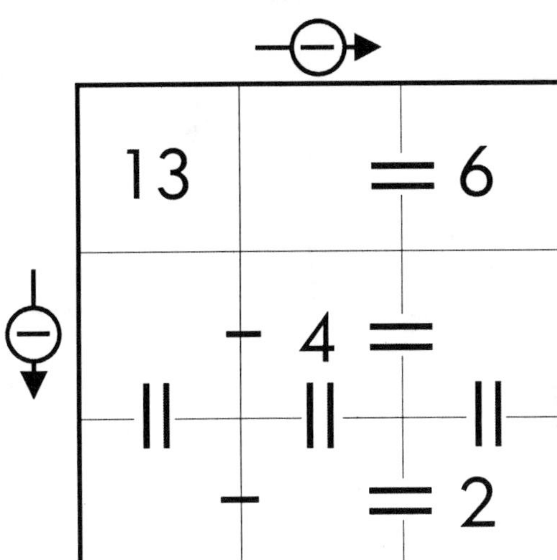

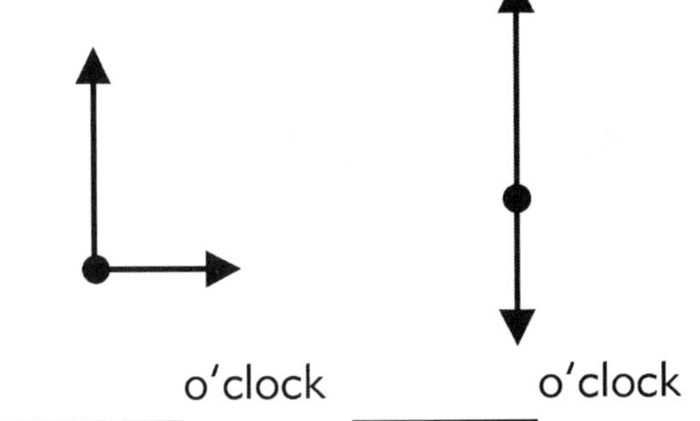

_______ o'clock _______ o'clock

2. Tell the time just by looking at the hands.

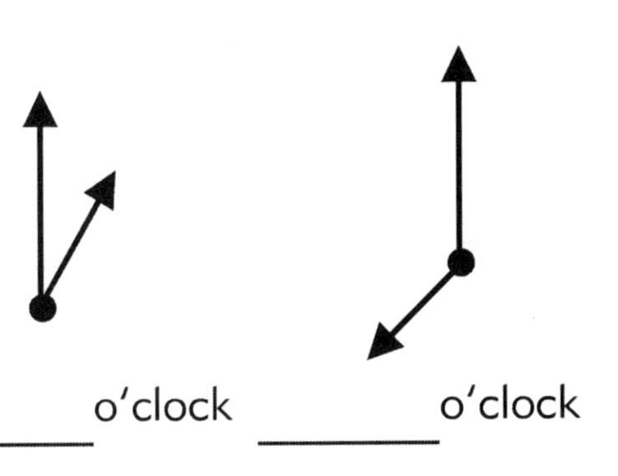

_______ o'clock _______ o'clock

3. Draw a line to match the design with its fraction.

1/3 1/2 ¼

1: Twenty-four

4. Use the code to get through the path. Write the numbers in each shape.

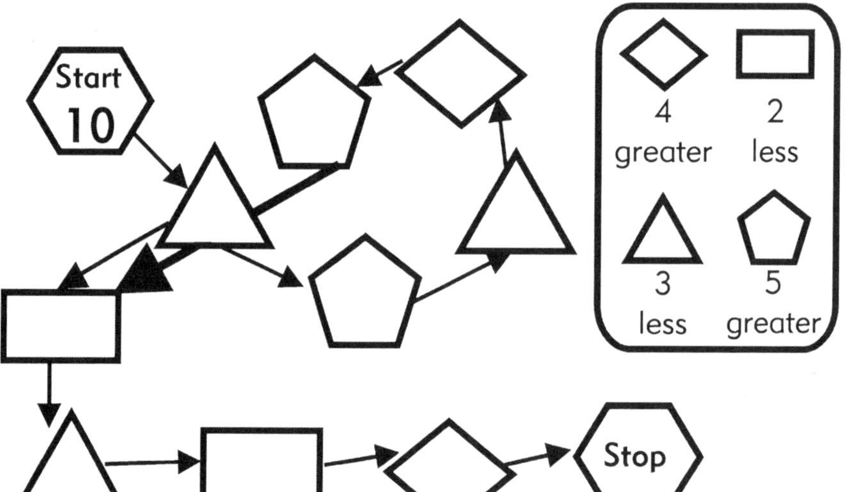

5. Crossnumber Puzzle. Across solves problems. Down checks problems.

ACROSS
1. 43 + 2
3. 55 + 1
5. 28 - 2
6. 24 + 3
7. 81 + 3
9. 29 + 2
10. 68 + 1
12. 18 - 1
14. 96 + 3
15. 19 + 2

1.	2.			3.	4.
5.				6.	
		7.	8.		
		9.			
10.	11.			12.	13.
14.				15.	

DOWN
1. 44 - 2
2. 54 + 2
3. 49 + 3
4. 68 - 1
7. 80 + 3
8. 42 - 1
10. 66 + 3
11. 97 + 2
12. 6 + 6
13. 73 - 1

6. Mike's score for 3 hits with darts was 61. He hit 14 and 33. On what number did his 3rd dart land?

Myself Partners Group

Math Rules!
1: Twenty-five

Name ______________________

1. Work to find 4 different ways Raj can wear his and .

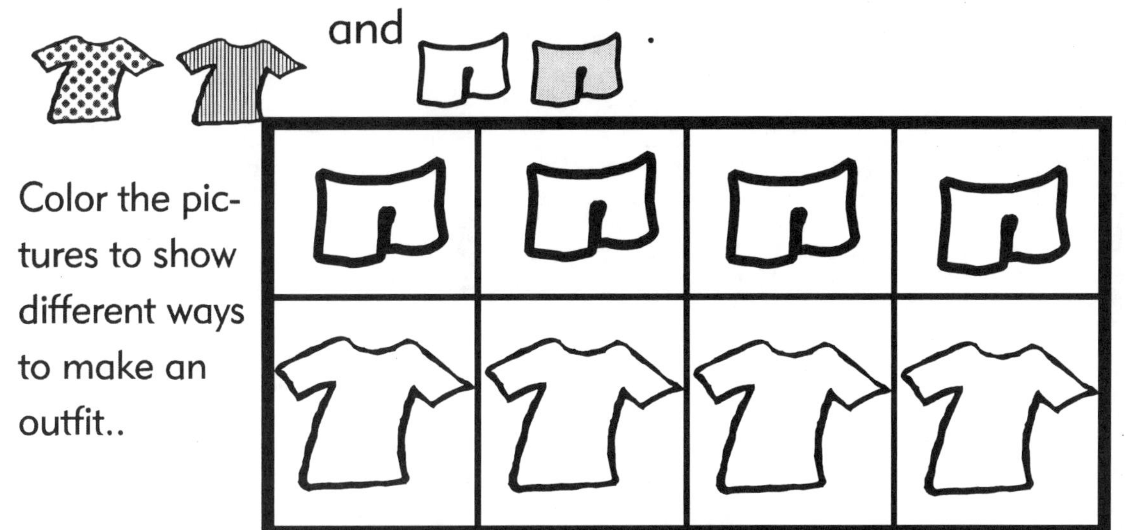

Color the pictures to show different ways to make an outfit..

2. Look at the calendar and circle the best answers.

a. Mother's Day is the second Sunday. 7 14 21

b. The last day of May in 2000 is a MONDAY TUESDAY WEDNESDAY

c. Memorial Day is the last day of the month.
27 28 31

MAY 2000

S	M	T	W	T	F	S
	1	2	3	4	5	6
7	8	9	10	11	12	13
14	15	16	17	18	19	20
21	22	23	24	25	26	27
28	29	30	31			

3. Circle the set that shows 1/3 striped.

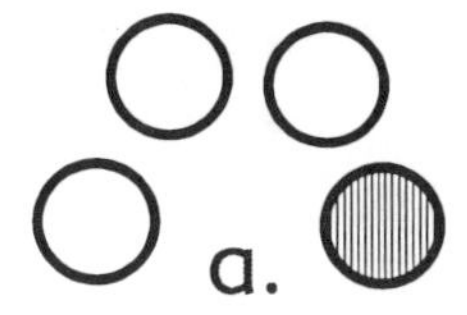

a.

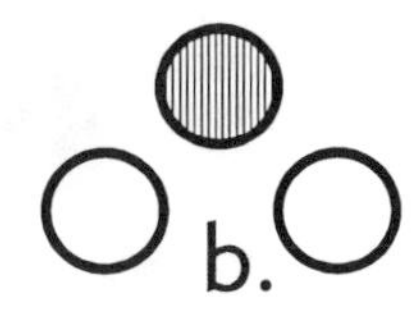

b.

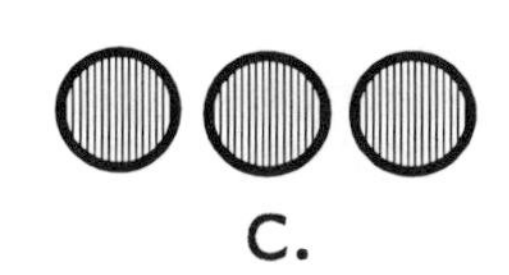

c.

1: Twenty-five

4. Find the secret number.

Start with 9
Subtract 3
Add 1
Double the sum.

The secret number is __________!

5. Help three friends share balloons. Make an equal set for each person.

6. Six marbles are in this bag. 4 are black and 2 are white. If you picked a marble without looking, what color are you more likely to get? Circle:

Black or White

Myself Partners Group

Second Grade

Name ______________________________

Math Rules!
2: One

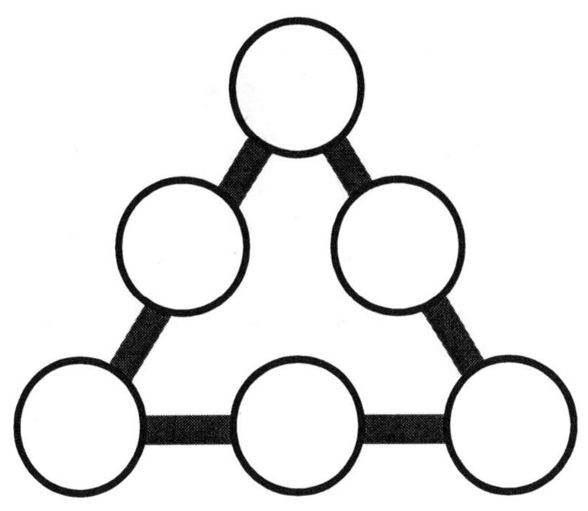

1. Fill in the circles using the digits 0 - 5. All three sides must add to 6.

2. Dustin had 6¢ in his pocket. His grandma gave him 3¢. He spent 5¢ on a gum ball. Then Dustin found a nickel, heads-up.

How much money does Dustin have now? __________

3. If a spider spins a web and traps 2 insects, how many legs are on the web?

____

4. Add

1+ 2 + 3 + 4 + 5 + 6 + 7 + 8 + 9 = ☐

5. Balance the scales. Write what's missing.

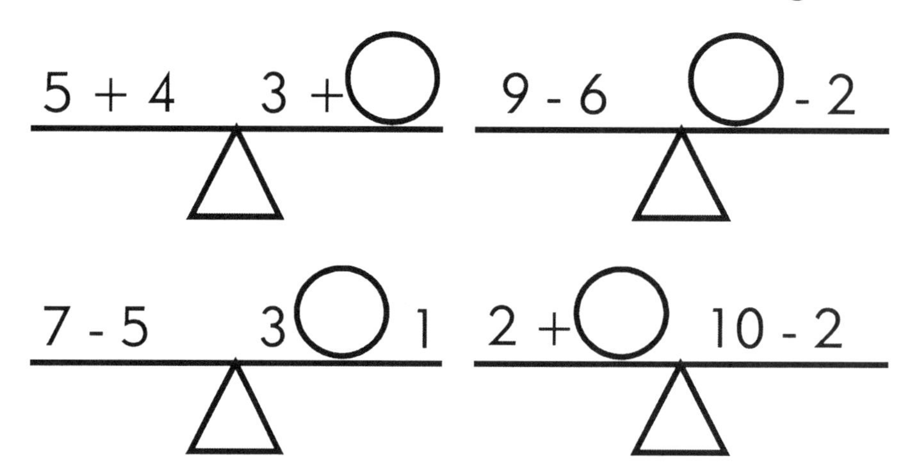

6. Fill in the blanks in this story of frogs and logs.

Five frogs sat on a log. Two frogs jumped into the water. Now there are ______ frogs on the log. Three spotted frogs jumped on the log. Then there were ______. A fly flew by and one frog fell off. Now there are ______ frogs on the log.

7. Fill in the last shape to continue the pattern.

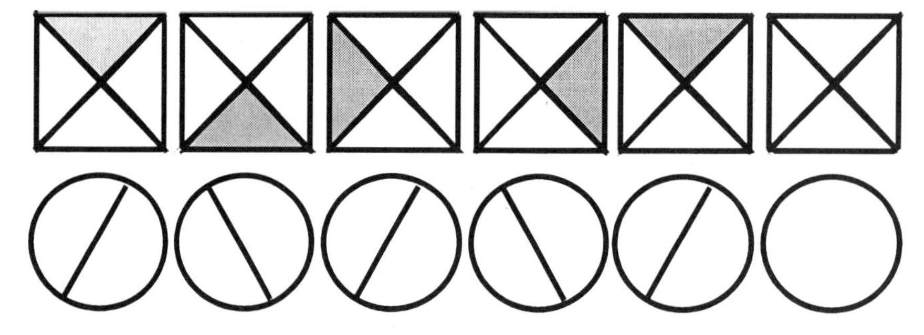

8. Choose the number that comes next.

2   5   8

a. 9   b. 10   c. 11   d. 12

Math Rules!
2: Two

Name ______________________

1. Write *all*, *some*, or *none*.

_____________ of the numbers are less than 20.

_____________ of the numbers are less than 10.

_____________ of the numbers are greater than 10.

_____________ of the numbers are greater than 20.

2. Tom gave names to his Matchbox® cars. He named the cars Mac, Bac, Tac, and Clack. Read the clues and match the names with the numbered cars.

Clues:

- Bac is next to Tac.
- Bac's number minus Tac's number equals Clack's number.
- Mac's number minus Bac's number equals Clack's number.
- Bac's number is not 7.

_______ MAC
_______ BAC
_______ TAC
_______ CLACK

3. Look at this number box. Fill in the missing numbers.

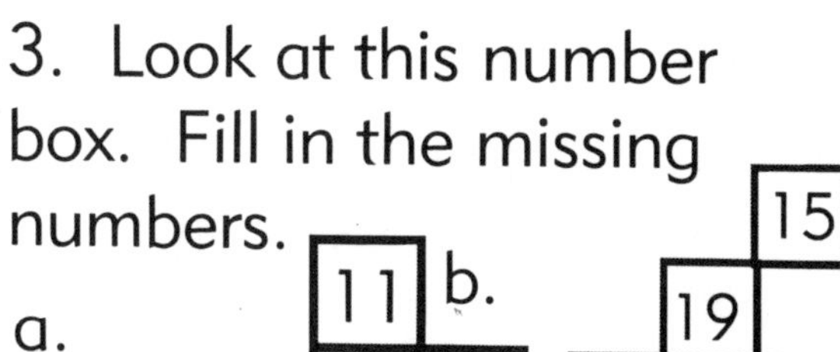
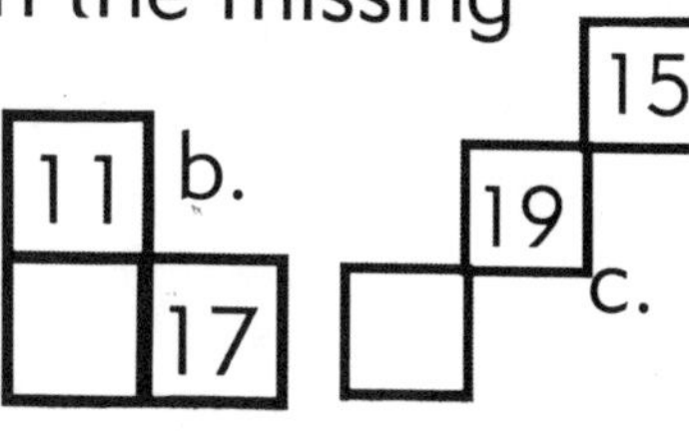

d.
13

1	2	3	4	5
6	7	8	9	10
11	12	13	14	15
16	17	18	19	20
21	22	23	24	25

2: Two

4. How many lines are used to make all the shapes in this pattern?__________ This shape is called ______________ .

octagon hexagon pentagon

5. Fill the missing numbers in the outside circle.

-7 -0 -1 -2 -3 -4 -5 -6 7

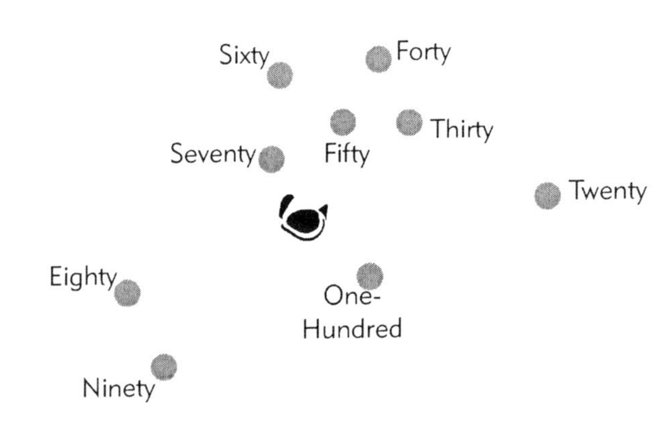

6. Connect the dots. Count by tens.

7. This spider can only move up or across toward the right to get to the fly. One path the spider took is shown. How many paths all together can the spider take to get to the fly? __________

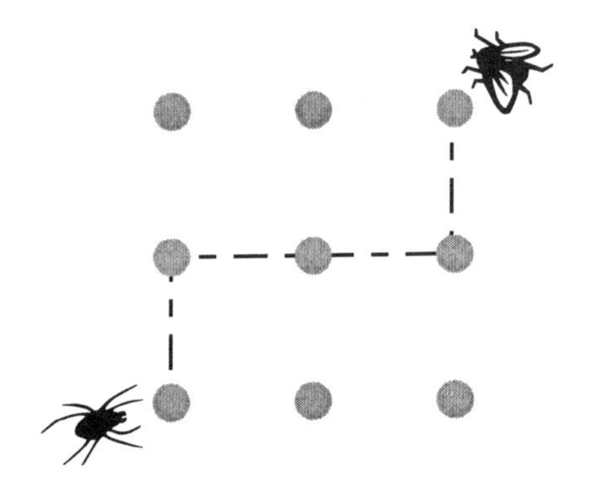

8. What number am I? I am an even number. Add me to 6 and you will get a number between 9 and 11. I am . . .

a. 2 b. 4 c. 5 d. 6

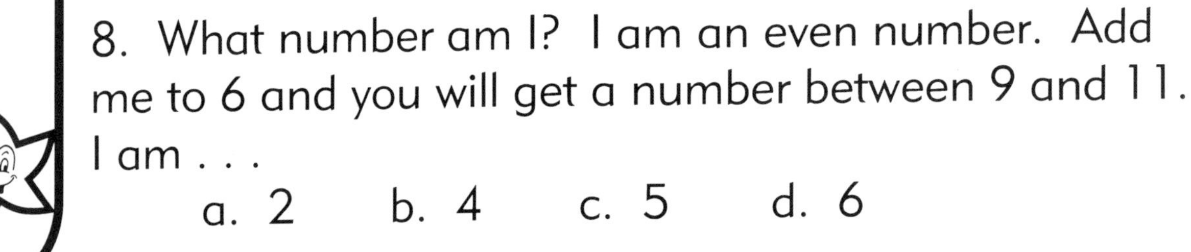

Math Rules!
2: Three

Name ______________________

1. Use the chart to answer these questions:

Number of Pages Read

Monday	◆ ◗
Tuesday	◗◗◗◗◗◗◗
Wednesday	◆◗ ◗ ◗
Thursday	◆ ◗◗◗◗◗◗◗
Friday	◆ ◗
Saturday	◆ ◗ ◗ ◗
Sunday	

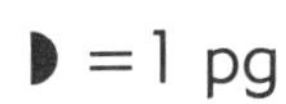 =1 pg ◆ =10 pgs

a. How many pages did Jeff read on Monday? __________

b. On which day did Jeff read 14 pages? __________

c. On which day did Jeff read the most pages?

____

d. Jeff read 12 pages on Sunday. Add this data on the chart.

2. Circle the items that round to 40.

a. 43 puzzle pieces b. 38 crayons c. 42 apples d. 47 sheets of paper

3. When a group of things can be put in pairs and <u>none</u> of them are left over, the number is called an even number.

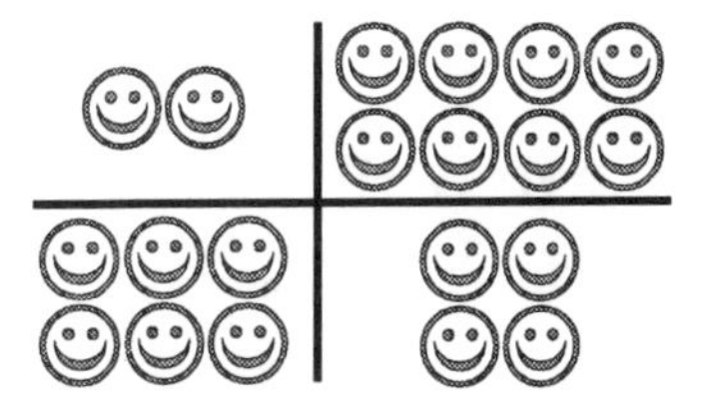

When a group of things can be put in pairs and <u>one</u> of them is left over, the number is called an odd number.

Circle if these numbers are odd or even.

13	Odd	Even
18	Odd	Even
25	Odd	Even
14	Odd	Even
19	Odd	Even

2: Three

4. Ancient Romans wrote numbers in a different way than you do.

Answer by using Roman Numerals:

Roman Numerals	I	II	III	IV	V	VI	VII	VIII	IX	X
Our Numbers	1	2	3	4	5	6	7	8	9	10

- fingers on your two hands _______ - wheels on a car_______
- your age _______ - players on a baseball team _______

5. These numbers are mixed up. Put them in order from smallest to largest.

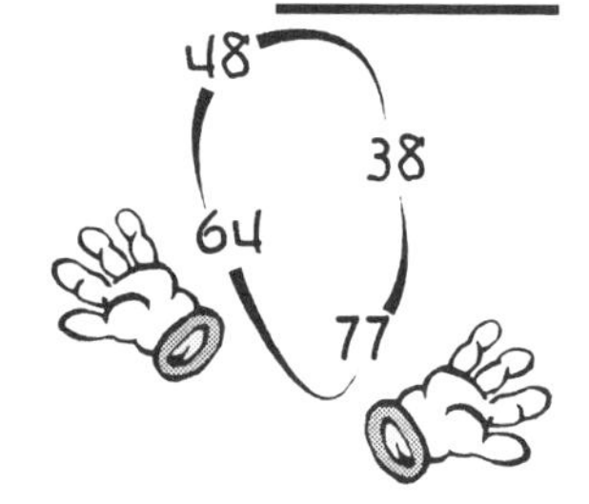

_______ _______ _______ _______

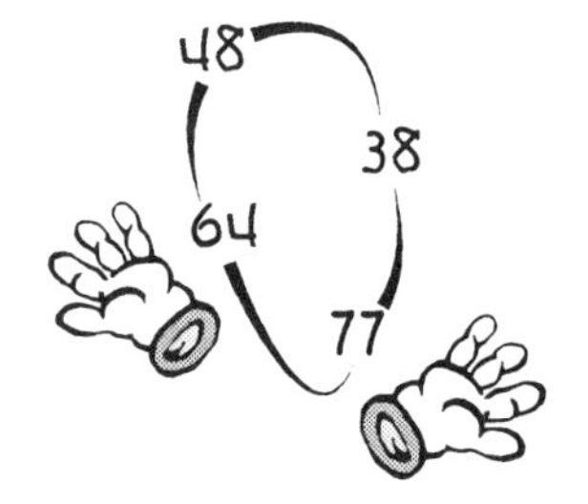

6. These numbers are mixed up. Put them in order from largest to smallest.

_______ _______ _______ _______

7. Color the map using your choice of four colors. NO state can be the same color as a state that touches it.

8. Alyssa saw that the seat she sat in at the movies did not have a number on it. The seat in front of her was 46. The seat behind her was 66. What was Alyssa's seat number?

a. 45 b. 47 c. 56 d. 65

Myself   Partners   Group

Math Rules!
2: Four

Name ____________________

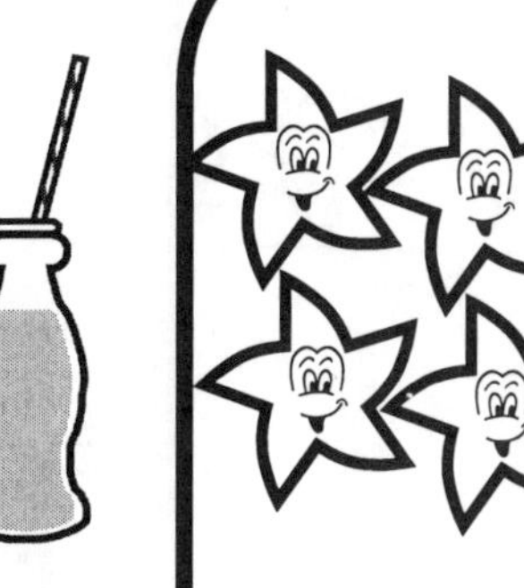

1. When Neil went to the bake sale his grandpa gave him a quarter. All the sweets looked good. He bought one of them and got 2 coins back as change. What did he buy?________________

11¢

35¢

45¢
22¢
19¢

2. Which weighs more -

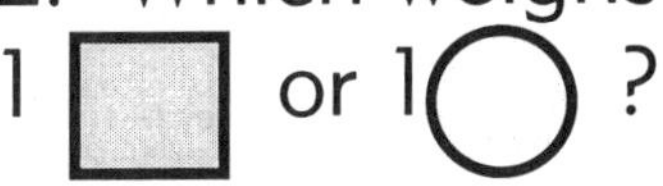

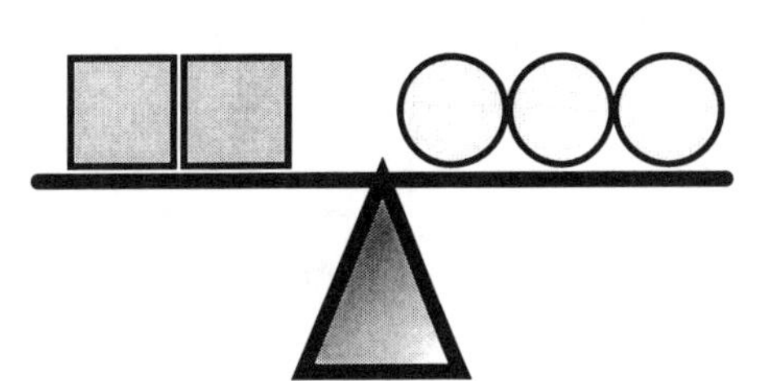

_________weighs more

3. There are six mailboxes next to each other across the street. The numbers begin at 27. The number on each mail box is 3 more than the number before it. Place the number on the last mailbox.

4. Underline your favorite amusement park ride. Circle the coins to show how much that ride costs.

You have

75¢-roller coaster / 55¢-ferris wheel / $1.25-merry-go-round

2: Four

5. How many apples hang on the tree?

____

6. Secret Code

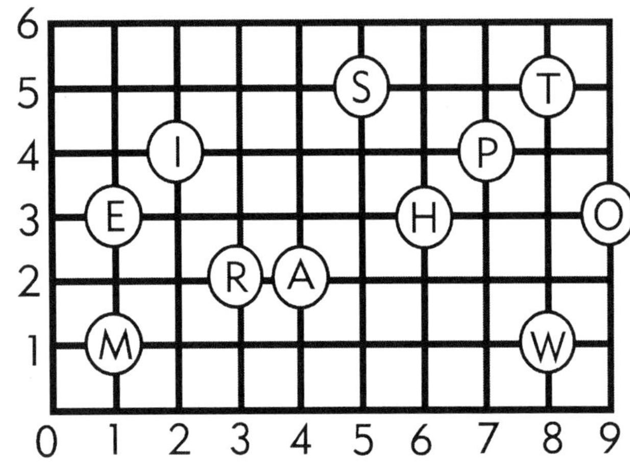

Move across from line to line, then up from line to line. Write the letter you find and read the message.

____ 1,1   ____ 4,2   ____ 8,5   ____ 6,3

____ 2,4   ____ 5,5

____ 7,4   ____ 9,3   ____ 8,1   ____ 1,3   ____ 3,2

7. Write the numbers that come between
48,_____ , _____ , 51

8. Seven people sat on the bus. They could not all sit in pairs. Then five more people got on the bus. How many pairs could there be now?
a. 6   b. 8   c. 10   d. 12

Myself Partners Group

Math Rules!
2: Five

Name ______________________________

1. Mayan Numbers. Look at these ancient numbers.

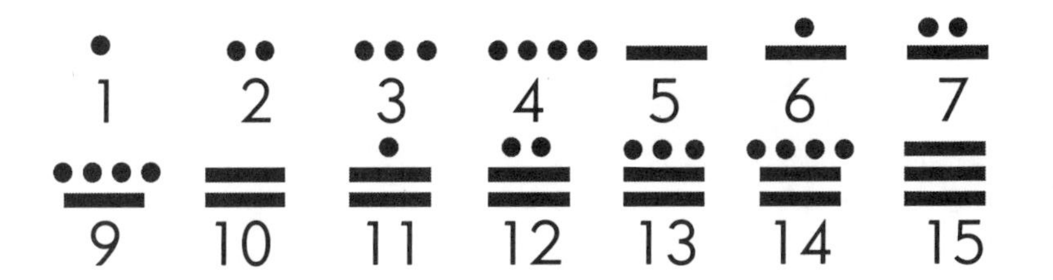

Write the Mayan numbers for

16 ________ 17________ 18 ________ 19 ________

2. Use a calculator to do these problems.

a. 26 + 53 + 174 + 309 = ____________

b. 674 - 59 + 481 = ____________

c. $1.37 + $3.49 + $ 5.15 + $ 0.28 = ____________

3. There are 8 pencils in a box. If your teacher asks you to give one pencil to each of the 22 students in your class, how many boxes do you have to get from the supply shelf? ____________

4. If there are 24 seeds in the large slice of water-melon, about how many seeds are in the smaller slice? ____________

5. A second grader always bought a fresh pretzel at lunch time. One pretzel costs 55 cents. About how much does he pay for a whole week's worth of pretzels? Circle your best choice.

$1 $2 $3 $4

6. How many chipmunks and squirrels are there all together?___ Count the number of feathered animals.___ If you arrange the ✦ so that all animal life is equal, how many animals would there be of each?___

Birds	✦✦✦✦✦
Squirrels	✦✦
Butterflies	✦✦✦✦
Chipmunks	✦

Each ✦ = 2

7. Greater than or less than.
Write > or < in the square for these Mayan problems.

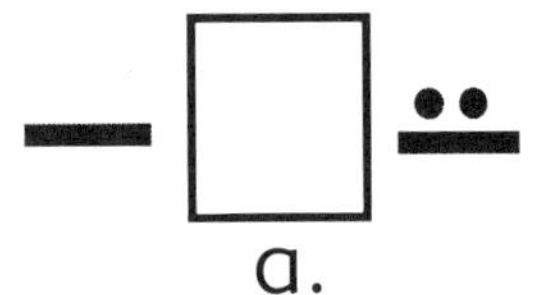

a.

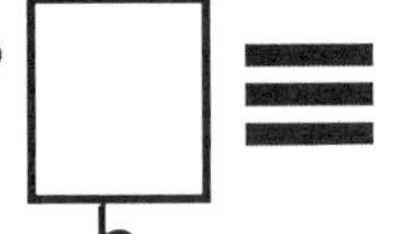

b.

c.

8. Julie is making ten beaded bracelets. She needs four red beads to add to other colors to make her pattern for each bracelet. Which bag or bags of red beads will she buy?

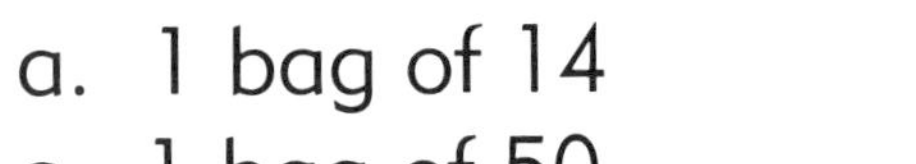

a. 1 bag of 14
b. 2 bags of 20
c. 1 bag of 50
d. 2 bags of 25

Math Rules!
2: Six

Name ______________________

1. Count all the squares in this design

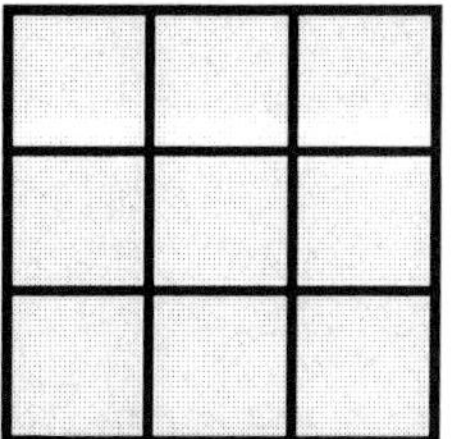

There are ____________ .

2. Use any 3 of these numbers to make correct number sentences. 2 3 4 6

EXAMPLE: [3] + [2] - [4] = 1

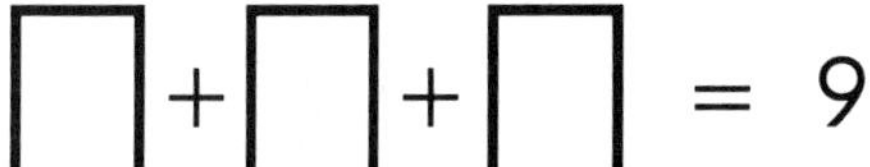

[ ] + [ ] + [ ] = 9

[ ] + [ ] - [ ] = 8

[ ] + [ ] - [ ] = 7

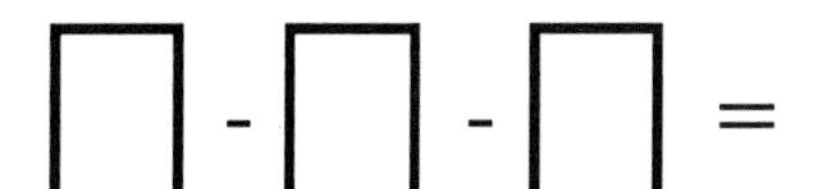

[ ] - [ ] - [ ] = 0

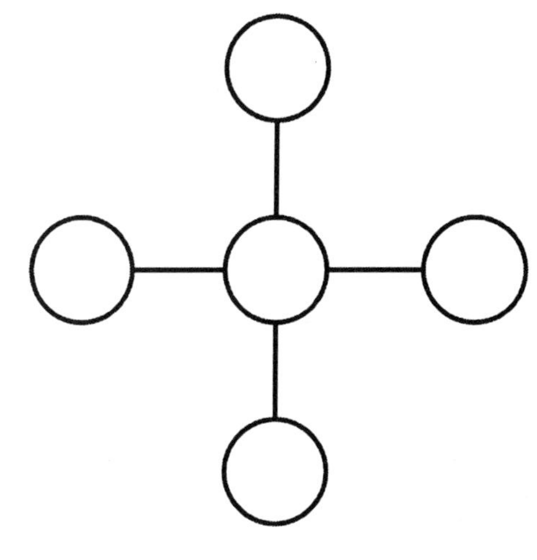

3. Put the digits 1, 2, 3, 4, 5 on this puzzle so that all lines have a sum of NINE.

4. A right angle looks like

Draw a line that makes a right angle at the point on the line segment.

2: Six

5. Barry bought a few supplies that he needed for school. He bought 3 pencils for 14¢ each, a pencil sharpener for 45¢, 2 large erasers for 39¢ each.

On the tax table, circle the amount of tax Barry paid on the total sale.

Tax Collection Schedule	
.01 to .15	.00
.16 to .17	.01
.18 to .34	.02
.35 to .50	.03
.51 to .67	.04
.68 to .83	.05
.84 to 1.00	.06
1.01 to 1.17	.07
1.18 to 1.34	.08
1.35 to 1.50	.09
1.51 to 1.67	.10
1.68 to 1.83	.11

6. Write the next numbers in this pattern.

19, 28, 37, 46, ____ ____ ____ ____ ____

7. The numbers in the pattern above have something in common. What do they have in common?

____

____

8. When the first day of June is a Thursday, on what day of the week does June 16 fall?

a. Thursday   b. Friday   c. Saturday   d. Sunday

Myself Partners Group

Math Rules!
2: Seven

Name ____________________

1. Draw the next set of dots in this pattern.

2. Calendar Capers FIND THE SPECIAL NUMBER.

JANUARY

S	M	T	W	T	F	S
					1	2
3	4	5	6	7	8	9
10	11	12	13	14	15	16
17	18	19	20	21	22	23
24	25	26	27	28	29	30

Begin on New Year's Day and follow the clues.
-Move ahead the number of days in a week.
-Move ahead the number of months in a year.
-Move back the number of letters in the sixth month.

The special number is __________.
The day of the week is __________.

3. Write the time and draw the hands on the clock.

20 min. before 4:30 3 hours and 10 min. after 2:15

__________ 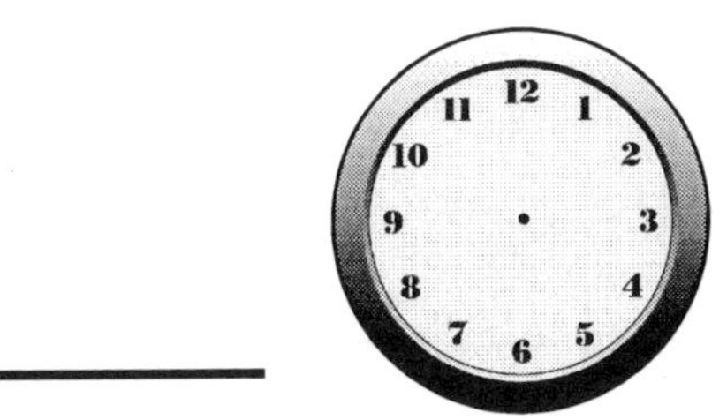 __________ 

4. Write the number sentence.

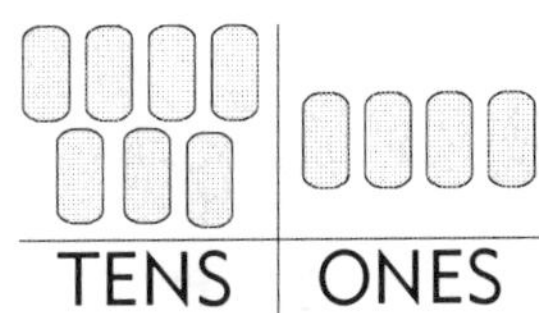

< >

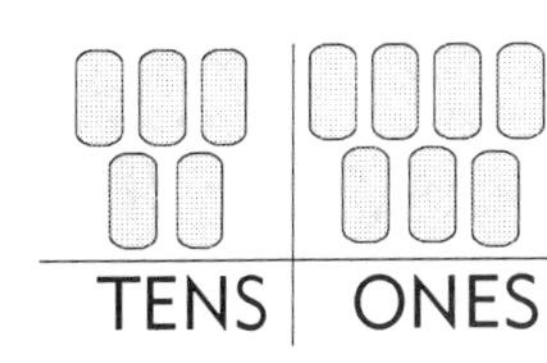

__________ ___ __________

2: Seven

5. First, ask 20 people what their favorite color is. Next, tally each response.

COLOR		TOTAL
red	______________________	
blue	______________________	
green	______________________	
yellow	______________________	
other	______________________	

Then, use the tally totals to make a pictograph.

Red	
Blue	
Green	
Yellow	
Other	

= 2 people (= 1 person)

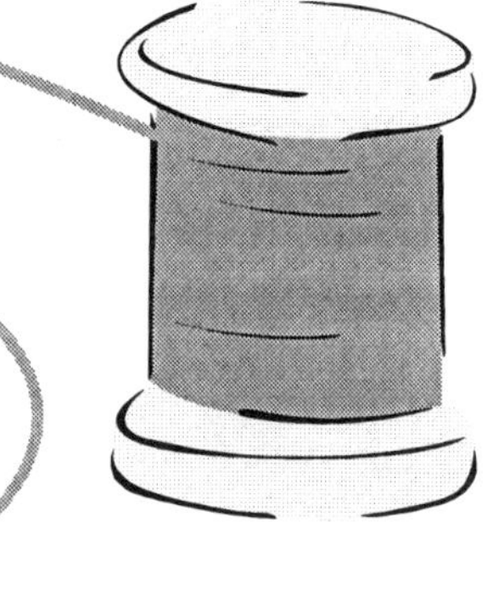

6. How many centimeters long is the piece of thread that is unwound from the spool? ______________

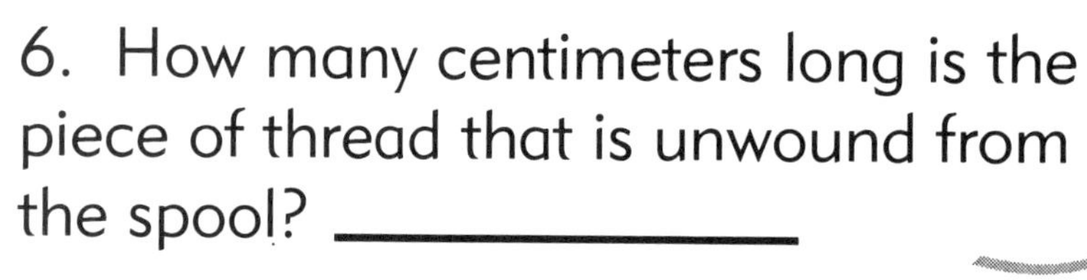

7. How many squares are in this design? ________________

8. Which letter of the alphabet is the fourteenth?
a. N b. O c. P d. Q

Math Rules!
2: Eight

Name ______________________

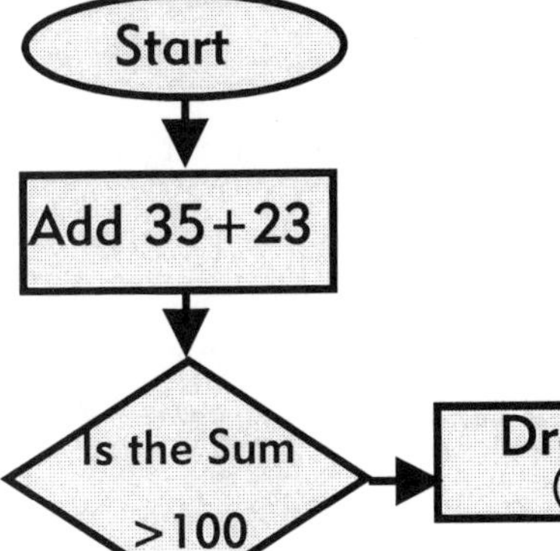

1. FLOWCHART.
Follow the steps.

Answer: ______________

2. Use your centimeter ruler to measure each part of this hand.

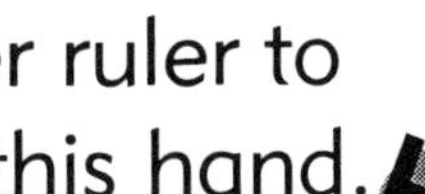

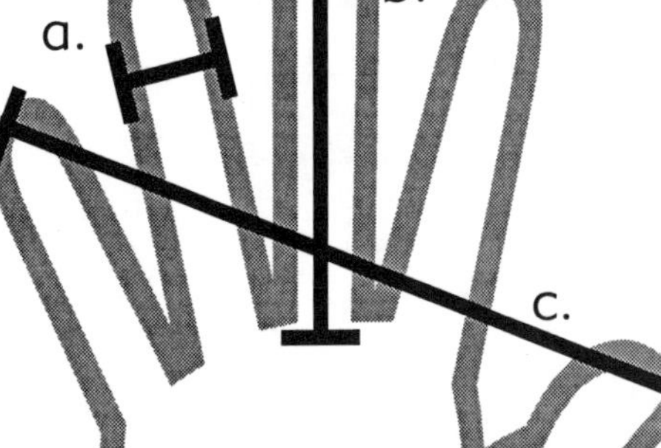

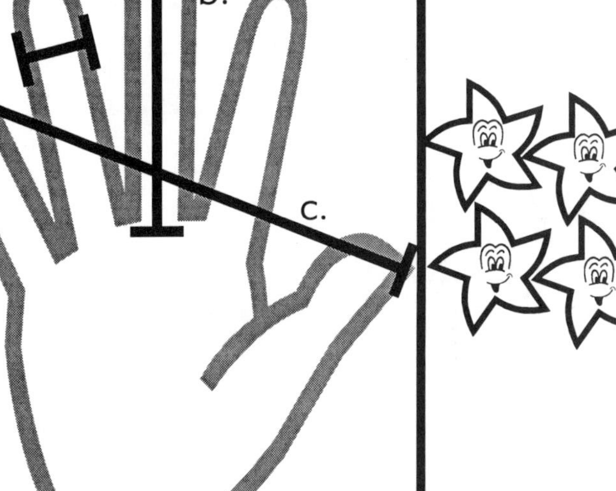

a. finger width ____________

b. finger length ____________

c. pinky to thumb ____________

d. wrist width ____________

3. Cross out one number in one of the triangles and write it in another triangle so that the sums of the digits in all three triangles are equal.

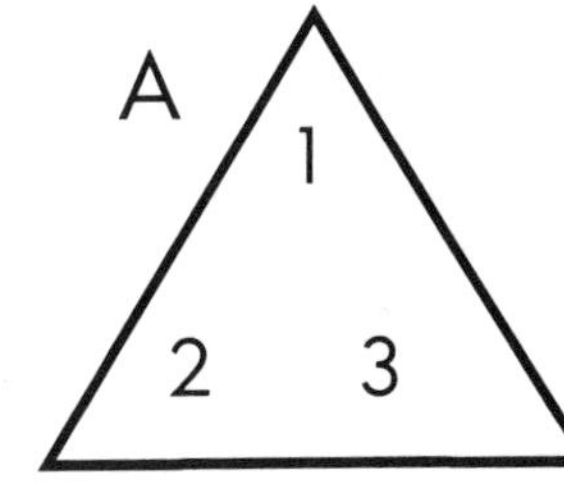

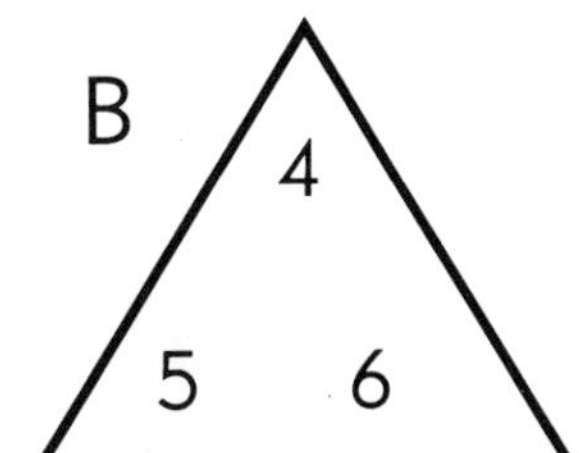

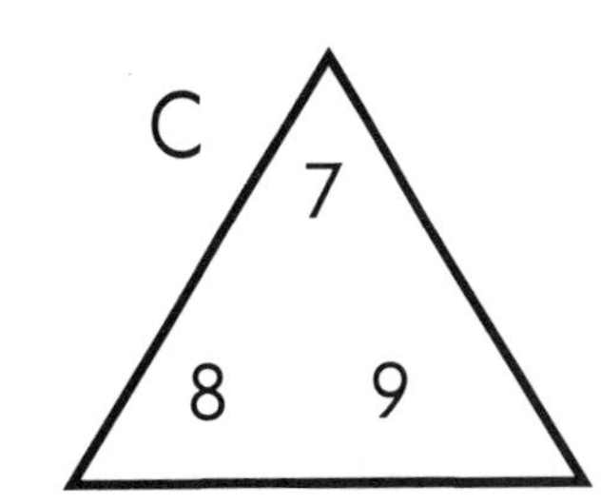

2: Eight

4. Fill in the box to make the equation true.

56 - ☐ = 23   6 + ☐ = 30   ☐ - 16=24

5. Kimberlee and Tony found a treasure map in a bottle. Name the location of the places on the map.

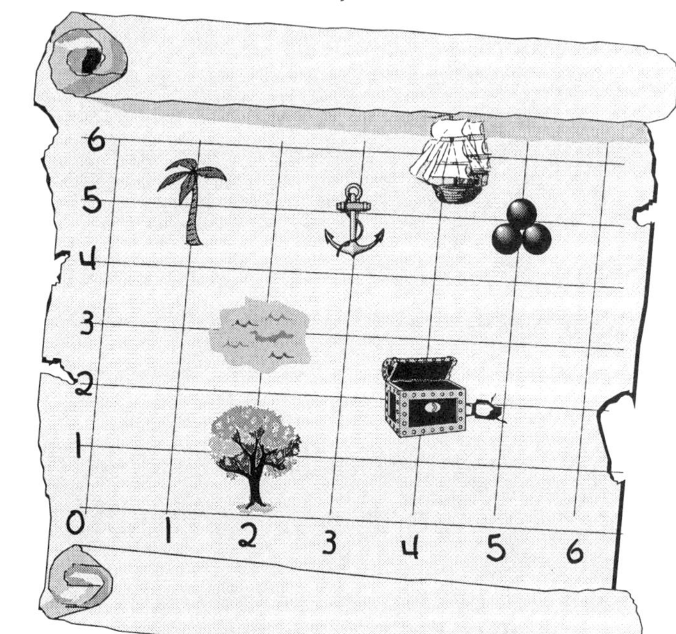

cannon balls

_______, _______

orange tree

_______, _______

pirate ship

_______, _______

treasure chest

_______, _______

6. Write the numbers that go in the squares.

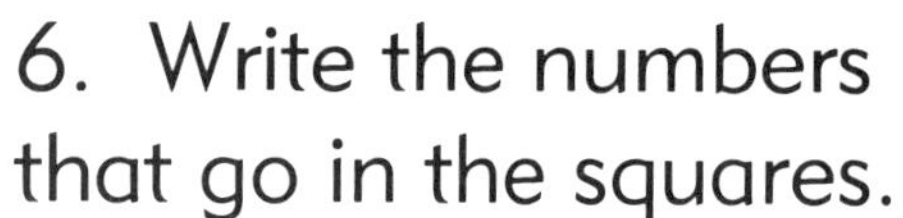

4	☐	7
−	8	☐
☐	4	4

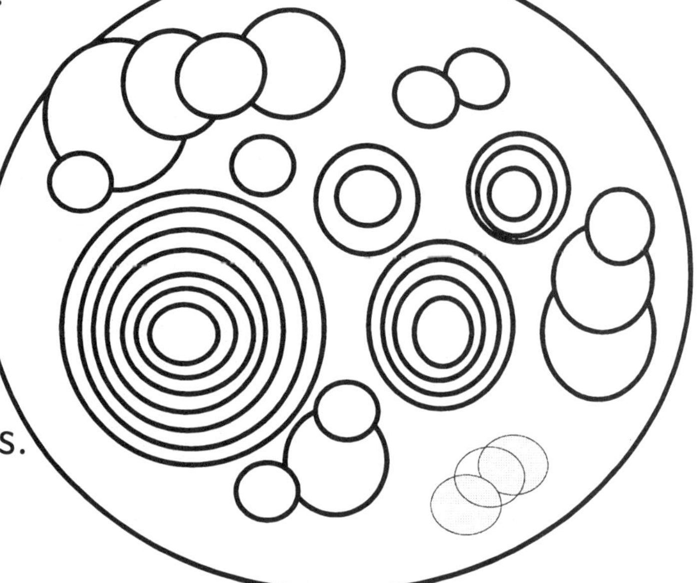

7. Count all the circles.
There are _______ circles in all.

8. Becky had a red T-shirt, yellow T-shirt, and a white T-shirt. She also had blue jeans, khaki pants, and a blue jean skirt. How many different outfits can she make?   a. 3   b. 6   c. 8   d. 9

Math Rules!
2: Nine

Name ____________________

1. Complete the table.

MONTH	ABBREVIATION	NUMBER
January	Jan.	first
March		
	May	
		ninth
November		

2. How many line segments can you make that connect the four dots?

____

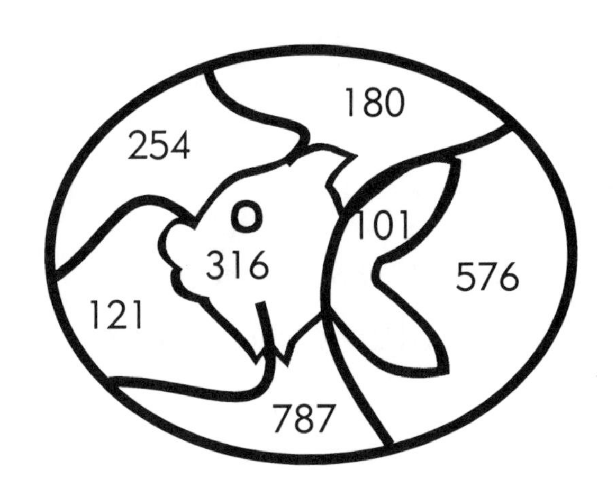

3. If the number has more tens than hundreds, color the area blue. If the number has more ones than tens, color the area yellow.

4. The AREA of a shape is the number of unit squares it takes to make the shape. Find the AREA of this shape using ☐ as the unit square.

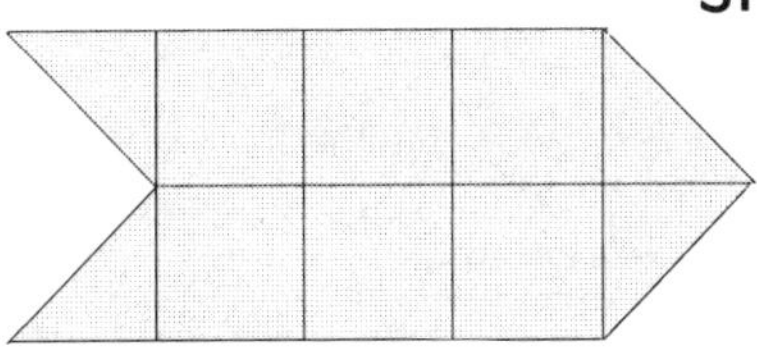

= __________ ☐s

2: Nine

5. Look to see how A compares to B. Now, which shape compares to C in the same way?

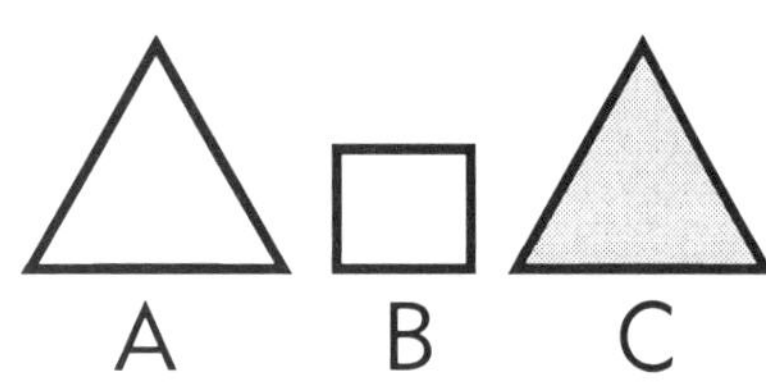

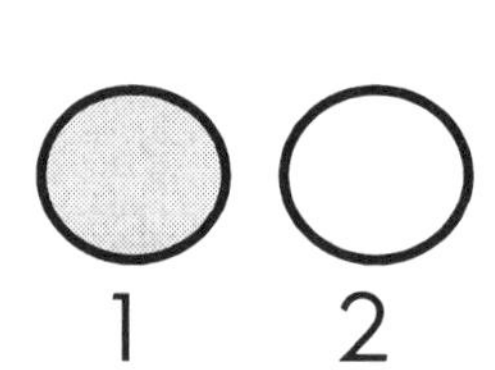

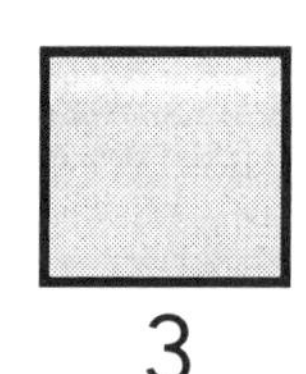

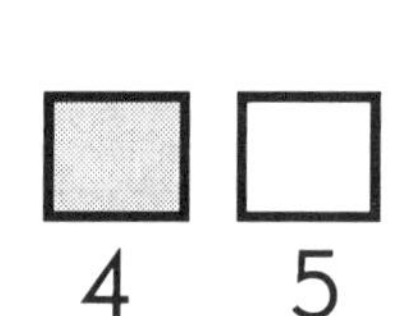

A B C 1 2 3 4 5

6. a. b. c.

Which purse has the most money? ____________

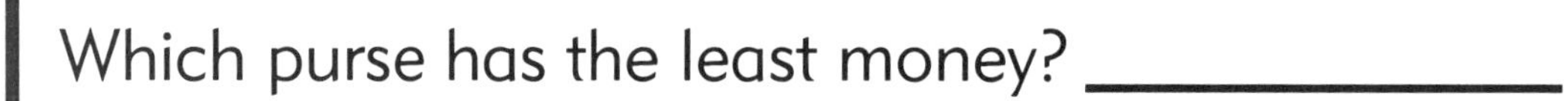

Which purse has the least money? ____________

7. How much money is in all three purses?

____

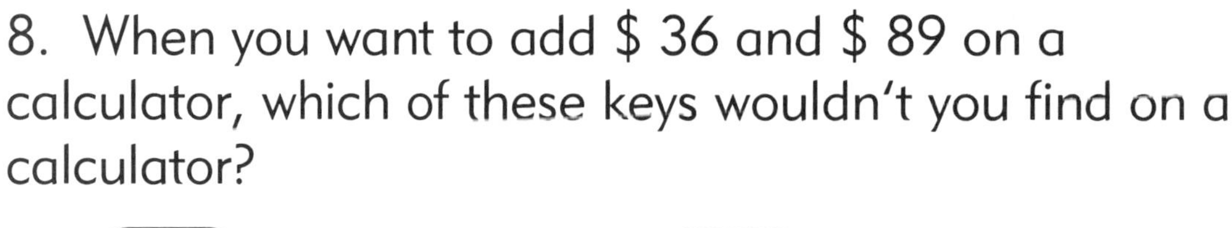

8. When you want to add \$ 36 and \$ 89 on a calculator, which of these keys wouldn't you find on a calculator?

a. 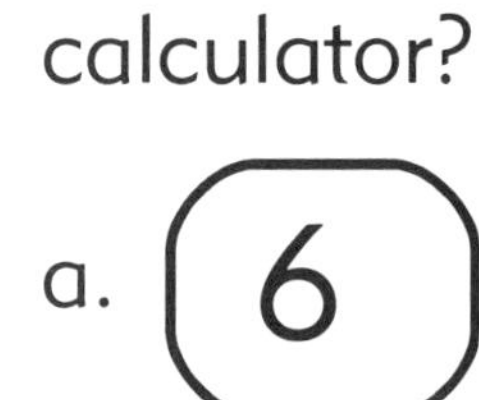

b. 8

c. 

d. \$

Myself Partners Group

Math Rules!
2: Ten

Name ____________________

1. The highest step in this set of stairs is 3 cubes high and it takes 6 cubes to make the stairs.
How many cubes will it take to make stairs where the highest step is 6 cubes high? ________ cubes

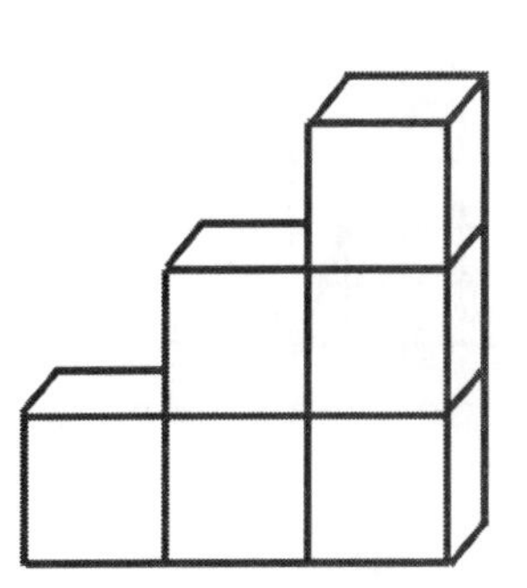

2. CLUES. What time did Cierra get home?________
* Cierra left school at 3:15 and walked to Chris house.
* It took Cierra 10 minutes to walk to Chris'.
* After 20 minutes and a snack, Cierra left Chris' house and walked 5 minutes to the library.
* Cierra returned 3 books and browsed for 1 5 minutes.
* Then she walked 10 minutes to get to her house.

3. Temperature is measured with a thermometer.
The unit for measuring temperature is called "degree."
Write the temperature that the thermometers show.

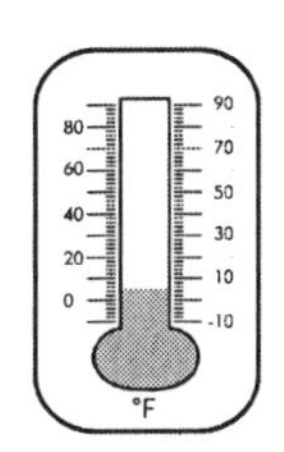
________ degrees
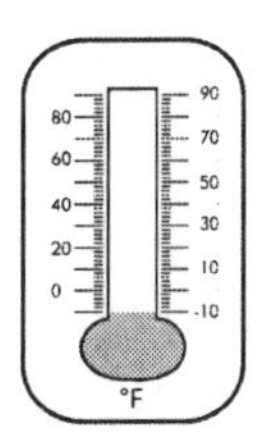
________ degrees
________ degrees

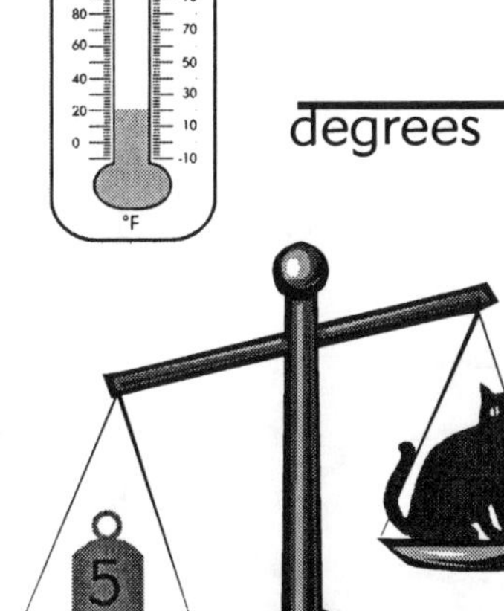

4. What is the MASS, or weight, of the cat?
a. 5 pounds b.less than 5 pounds
c. more than 5 pounds

5. Draw the same shape and size of these two designs.

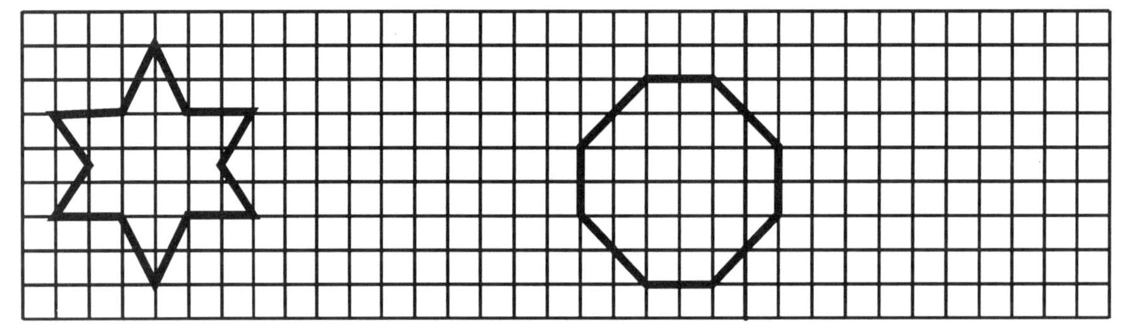

6. CODE:

⇧ move up

⇨ move to the right

⇩ move down

⇦ move to the left

Use the chart to fill in the numbers.

0	1	2	3	4	5	6	7	8	9
10	11	12	13	14	15	16	17	18	19
20	21	22	23	24	25	26	27	28	29
30	31	32	33	34	35	36	37	38	39
40	41	42	43	44	45	46	47	48	49
50	51	52	53	54	55	56	57	58	59

34 ⇨⇨⇧ = 26

a. 43 ⇨⇧⇨⇧ = ☐

c. 14 ⇨⇧⇦⇩ = ☐

b. 59 ⇧⇦⇧⇧⇨ = ☐

d. 53 ⇧⇧⇧⇧ = ☐

7. How many "triominoes" does it take to fill the grid?

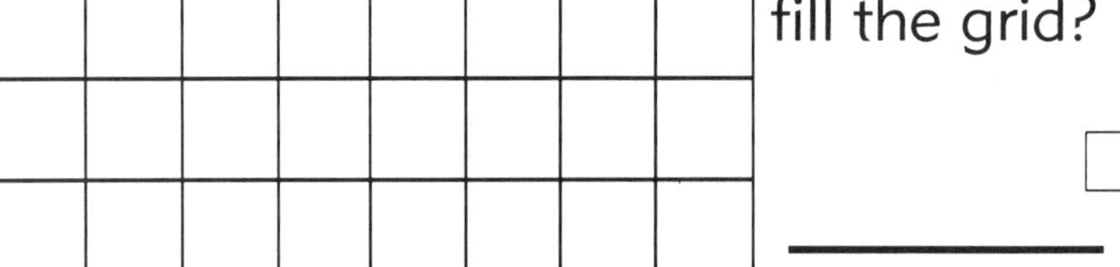

_______ s

8. Estimate the centimeters the ant walked from START to STOP when it took the bottom path.

START

94 cm

STOP

72 cm

a. 80 cm   b. 120 cm   c. 160 cm   d. 188 cm

Myself

Partners

Group

Math Rules!
2: Eleven

Name ______________________

1. Write the number that tells how many . . .

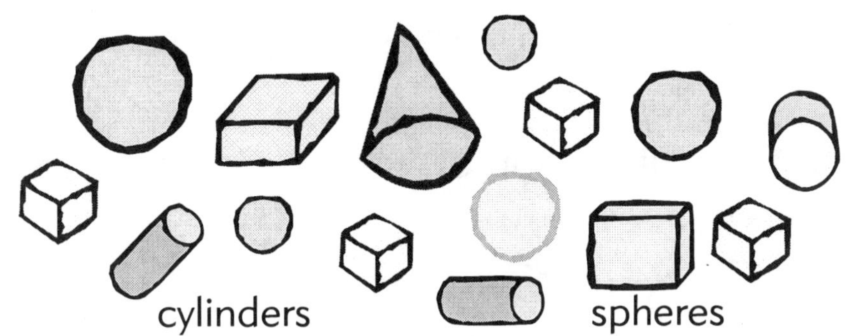

__________ cylinders

spheres __________

cones __________

__________ cubes

2. Put in order the letters representing the different areas from smallest to greatest.

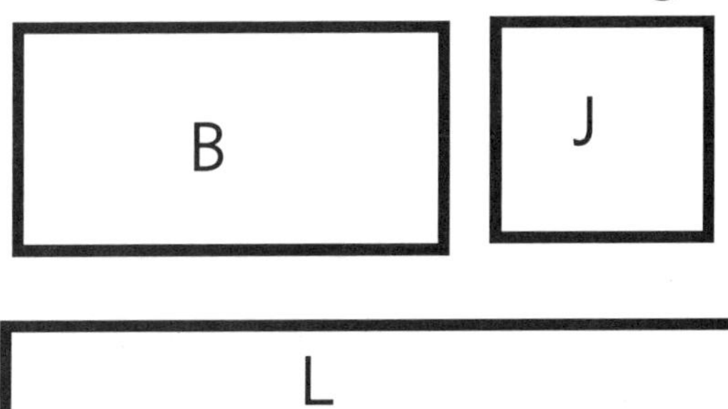

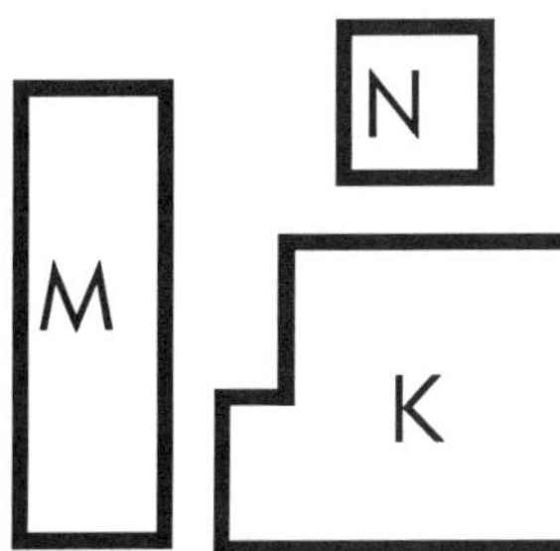

J

L

___ ___ ___ ___ ___ ___ ___

3. Use this chart to decode the letters in your answer from problem 2.

A	B	C	D	E	F	G	H	I	J	K	L	M	N	O	P	Q	R	S	T	U	V	W	X	Y	Z
I	G	S	O	A	D	T	Q	J	E	L	N	B	U	R	M	K	W	F	Y	P	X	H	V	Z	C

___ ___ ___ ___ ___ ___ ___

4. If you took 8 milk cartons from the counter at the cafeteria, how many milk cartons will you have? _____

2: Eleven

5. Count the triangles in this design. The triangles come in many different sizes.

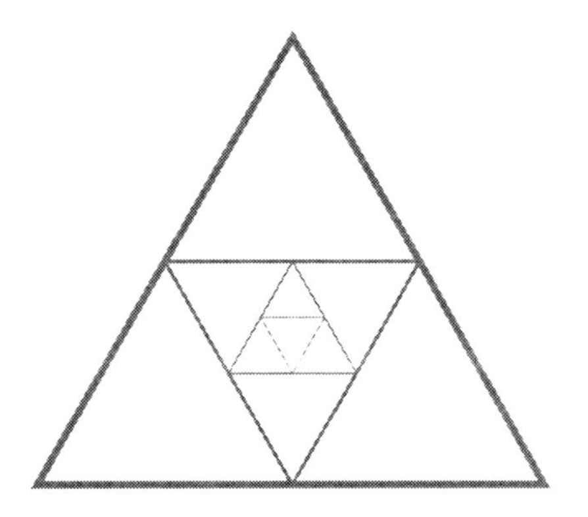

THERE ARE ____ TRIANGLES.

6. Use any combination of + and - to make true number sentences.

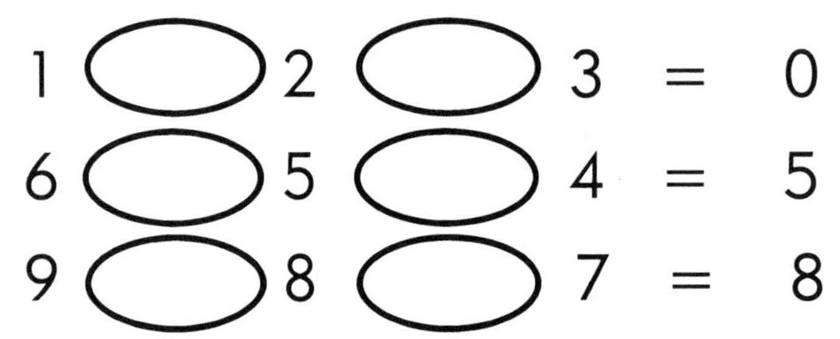

7. At the lunch table, 8 students drank one cup of milk each. Fill in the pictochart to show how much milk they drank in all.

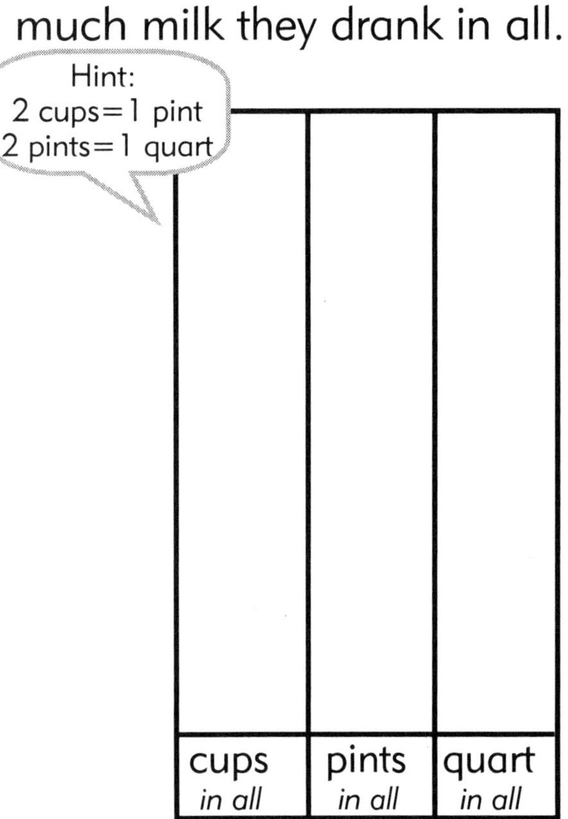

8. Daniel worked a problem on his calculator. When he turned the calculator upside down, he saw a word — See if you can find the problem he put into the calculator.

a. 335
-133

b. 3
3
+5

c. 111
112
+112

d. 35
+30

Myself

Partners

Group

Math Rules!
2: Twelve

Name ______________________

1. Check the correct column.

	Weight in ounces	> one pound	< one pound	= to 1 pound
	5 ounces			
	16 ounces			
	42 ounces			
	13 ounces			
	22 ounces			
	16 ounces			

2. Complete the subtraction wheel.

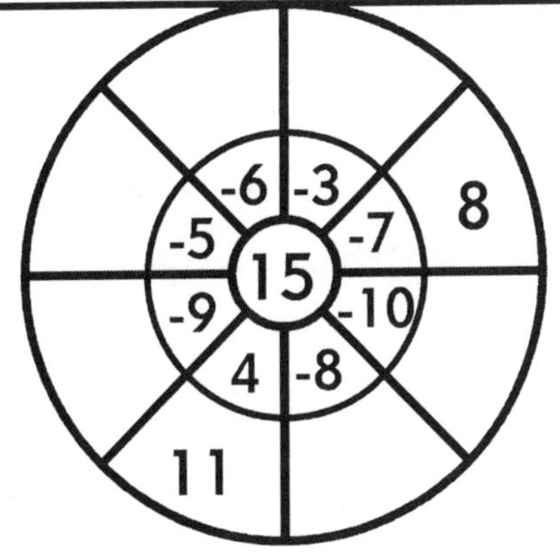

3. Chelsea had 14 collector marbles. She gave Zach 4 of them. Zach added them to his collection of 5 marbles. Now who has more marbles?

__________ has _____ more marble(s) than __________ .

4. Fill in the area that is missing in the fifth hexagon in this pattern.

5. Match the best number with these items.

Dollar _____	a. one
Bicycle _____	b. two
Days of the week _____	c. four
Quarts in a gallon _____	d. seven
Months in a year _____	e. ten
Dime _____	f. twelve
Quarter _____	g. twenty-five
Unicorn _____	h. one-hundred

6. Here are two groups of beads that Jill plans to make into two bracelets.

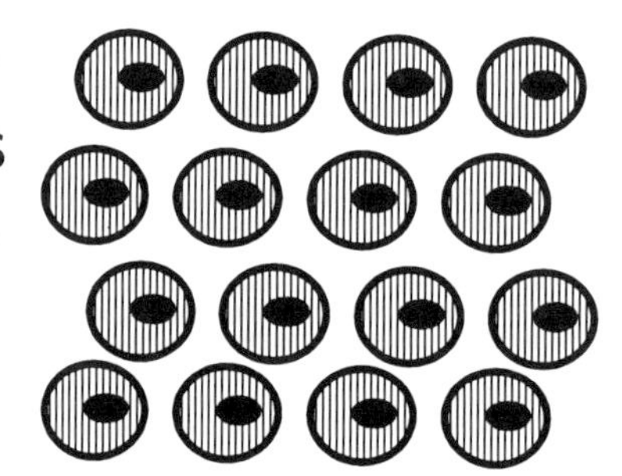

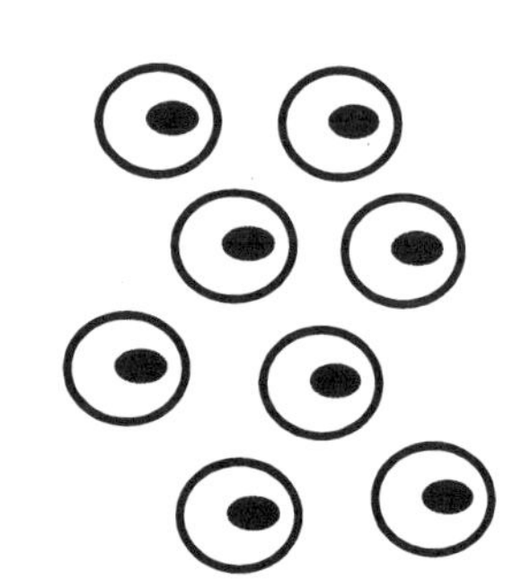

Without adding more beads or leaving any out, write a number sentence to show what she must do to make both groups of beads equal.

stripped _____ - _____ = _____

solid _____ + _____ = _____

7. Add or subtract to find the missing numbers.

✧ 2 + 3 + 4 + _________ = 16

✧ 7 + 8 - 9 + _________ = 15

✧ 4 + 4 + 4 + _________ = 16

8. A large bottle of cola is measured in

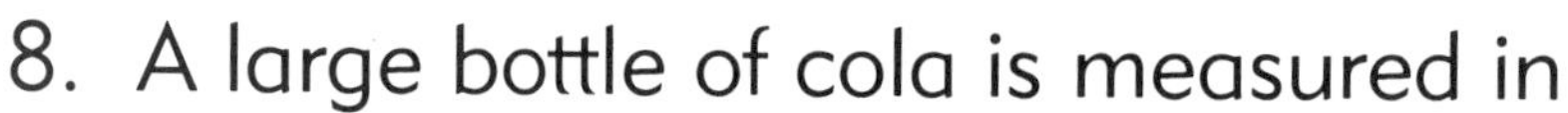

a. celsius b. grams c. liters d. meters

Math Rules!
2: Thirteen

Name ____________________

1. Circle the tools.

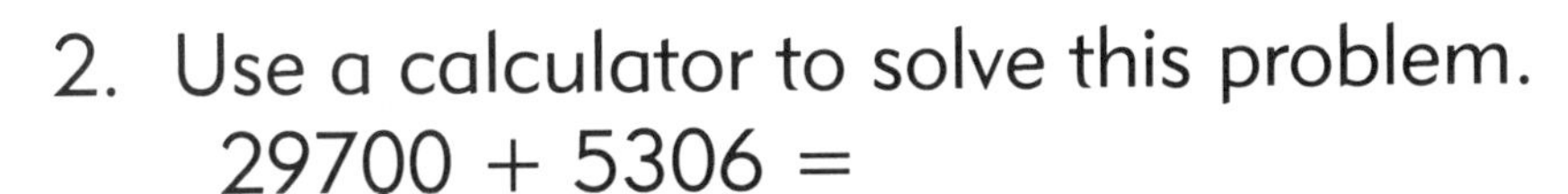

2. Use a calculator to solve this problem.
   29700 + 5306 = ____________

Turn the calculator upside down. What word does the sum spell? ____________

3. How many ways can you make 15 cents? Fill in the chart.

pennies	nickles	dimes
15	0	0

4. Circle the best answer..
   Marcus's cat drinks ________ of water each day.

less than a liter　　　a liter　　　more than a liter

2: Thirteen

5. The number of times a heart beats for one minute is called a pulse.

A person's pulse rate is about 72 beats per minute.

PULSE RATES OF ANIMALS

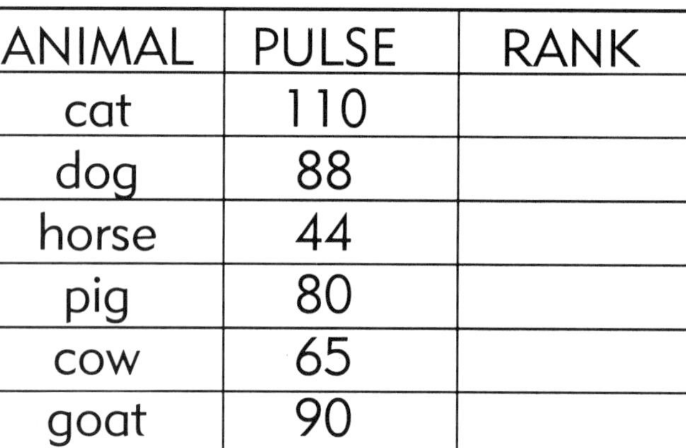

ANIMAL	PULSE	RANK
cat	110	
dog	88	
horse	44	
pig	80	
cow	65	
goat	90	

a. Which animal has the highest pulse rate? ________

b. Which animal has the lowest pulse rate? ________

c. Which animals have a pulse rate higher than a person's rate? ______________________________

6. Go back to the last column in the table in problem 5 to RANK (put in order) the pulse rates of the animals. Use the numbers 1 - 6 to rank from lowest to highest.

7. Skip-count by:

threes . . . . 35, 38, ____, ____, ____, ____

fours . . . . . 82, 86, ____, ____, ____, ____

8. A changes to B like C changes to? __________

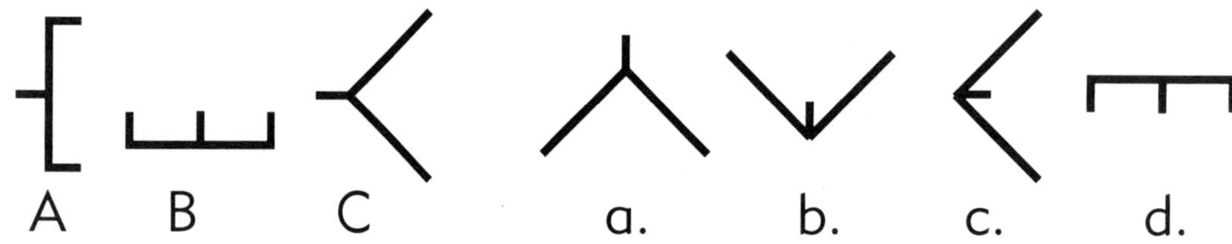

 CLC0239

Math Rules!
2: Fourteen

Myself Partners Group

Name ______________________

1. Write TRUE or FALSE beside each sentence.

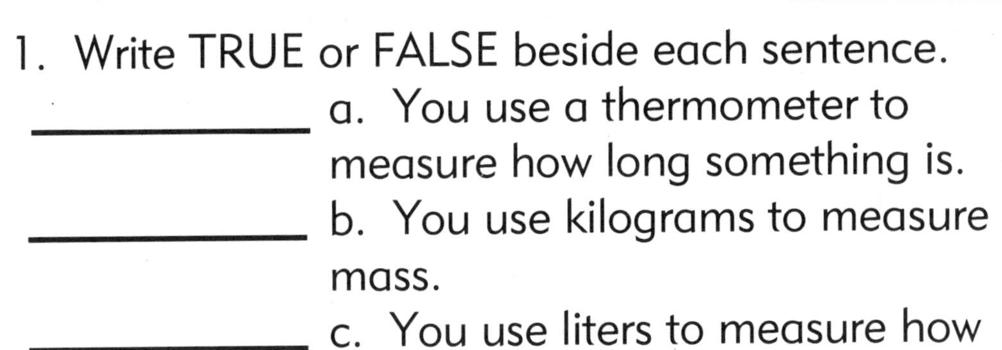

__________ a. You use a thermometer to measure how long something is.

__________ b. You use kilograms to measure mass.

__________ c. You use liters to measure how much a liquid weighs.

2. A. Mrs. Lady Bug is at the bank. She needs to go to the school to get her children. About how many blocks must she travel?

____

B. Bobby Bug is in school. Is it closer for him to walk past the tree or the bank to get home? About what is the difference?

____

C. About how far does Mr. Lady Bug have to go to get from Day Kare (where Baby Bug spent the afternoon) to home? __________

Day Kare

Equals about 1 block

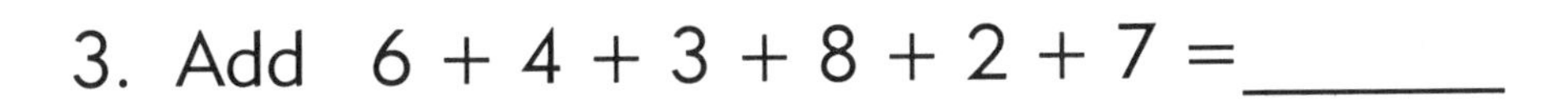

3. Add 6 + 4 + 3 + 8 + 2 + 7 =______

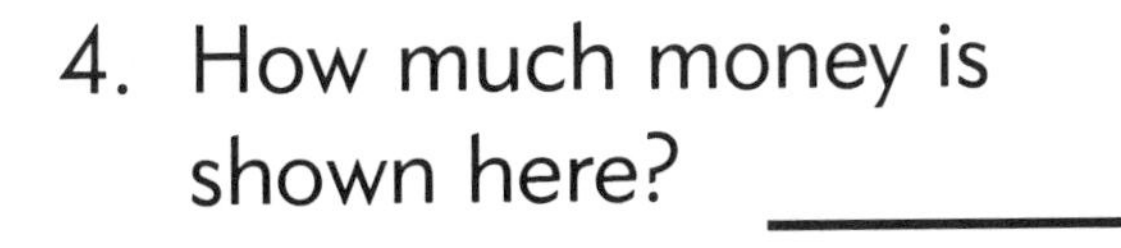

4. How much money is shown here? ________

5. DRAW A LINE TO THE PRICE OF ONE ITEM.

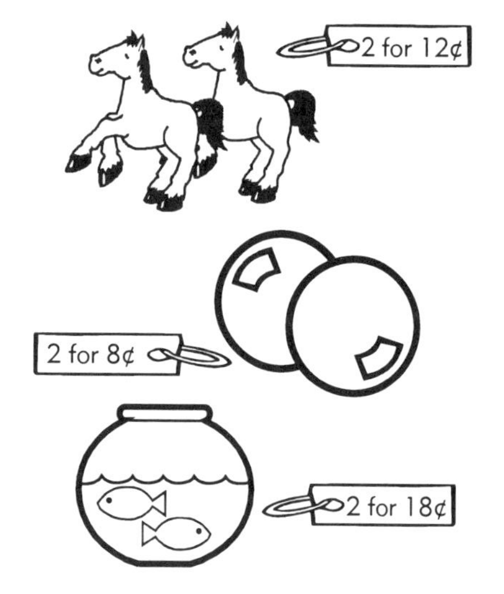

10¢
4¢
9¢
7¢
5¢
6¢
8¢

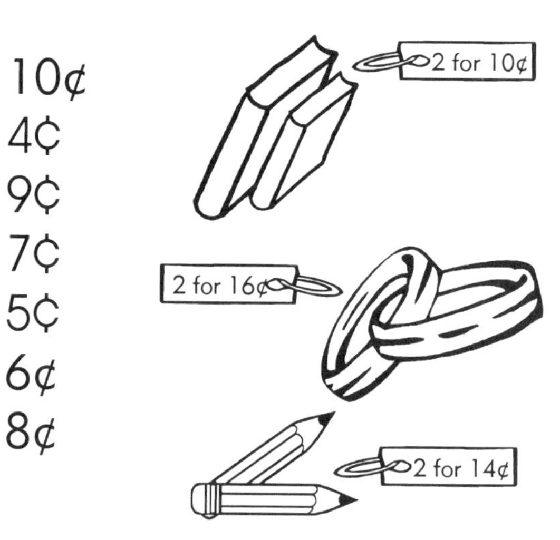

6. Jacob's shoe is twice as big as Austin's shoe. Jacob measured his desk at school and said it is 4 of his shoes long. How long is the desk if Austin measured it with his shoe? ______________

7. Temperature is measured in degrees. Write the temperatures you see on the four thermometers.

____°F

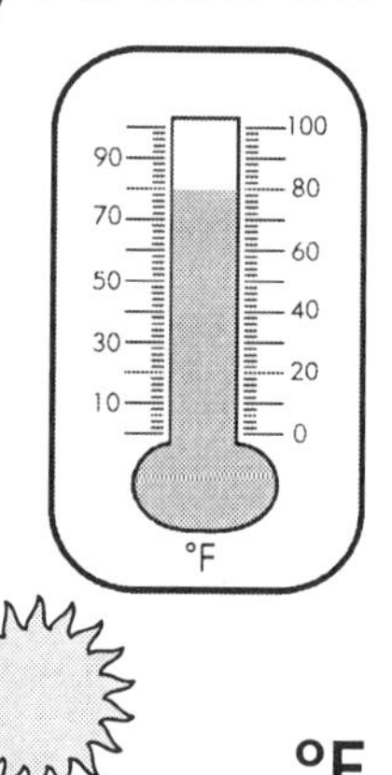

____°F

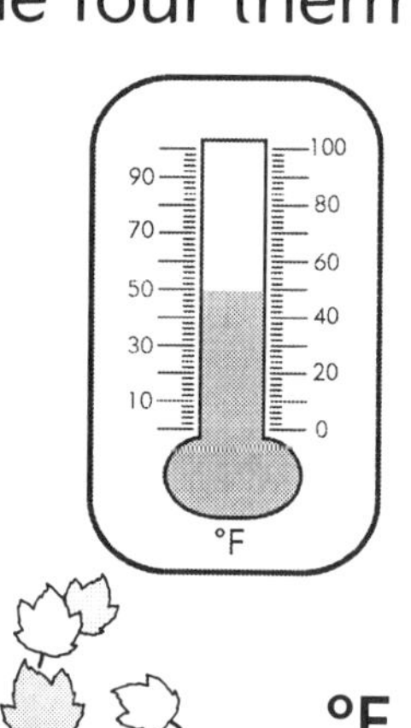

____°F

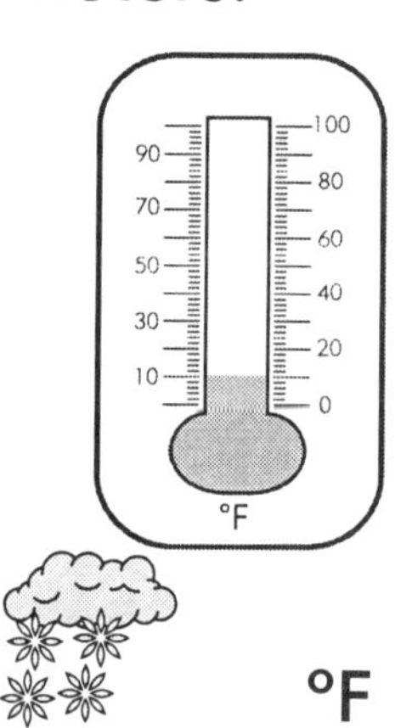

____°F

8. How many minutes are in 3 1/2 hours?

a. 150 minutes b. 120 minutes c. 90 minutes d. 210 minutes

Math Rules!
2: Fifteen

Name ______________________________

1. The distance all the way around a shape is called the perimeter. Use a centimeter ruler to measure the perimeter of this triangle. The perimeter of the triangle above is ____cm.

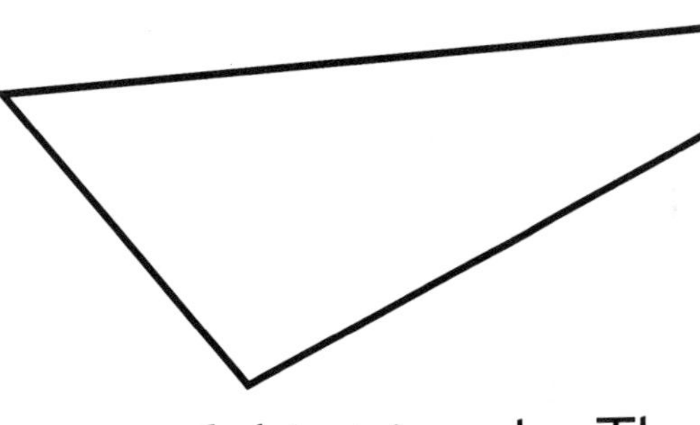

2. A RIGHT ANGLE looks like this paper 
The corner of a piece of paper fits snuggly into it. Use a regular piece of paper to decide which of these angles are right angles. Then circle them.

A B C D

3. Color the fourth from the left strawberry red. Color the twelfth strawberry green. Color the nineteenth strawberry brown.

4. Write the standard numeral.

a. 60 + 7 + 500 = ______
b. 3 + 800 + 20 = ______
c. 100 + 4 + 30 = ______

2: Fifteen

5. Discover the number.
I am an odd number between 6 and 11.
I am not the sum of 3 + 4.
What number am I? ______

6. Mike has 3 goldfish. One of the goldfish had 8 babies. Mike's dad let him keep only 2 babies.
How many must he give away? ________

How many goldfish does Mike have now? __________

7. Solve the coded number sentences.

0	1	2	3	4	5	6	7	8	9	10	11	12
A	B	C	D	E	F	G	H	I	J	K	L	M

1. G + G = M
2. L - D =
3. J + C =
4. B D - M = B
5. K - H =
6. J + L =

8. How many even numbers between 20 and 50 have a 3 in them?

a. 10 b. 5 c. 20 d. 15

Math Rules!
2: Sixteen

Name ______________________

1. This spinner can stop on any of the four spaces. Each space is the same size but a different pattern. The chance or PROBABILITY of the spinner stopping on the dotted space is 1 out of 4. Probability is often written as a fraction, ¼. What is the probability the spinner will stop on the dotted space?

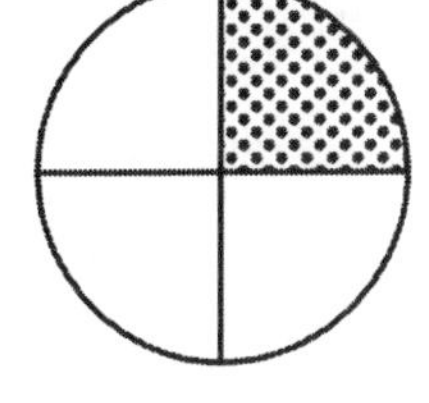

______ ______

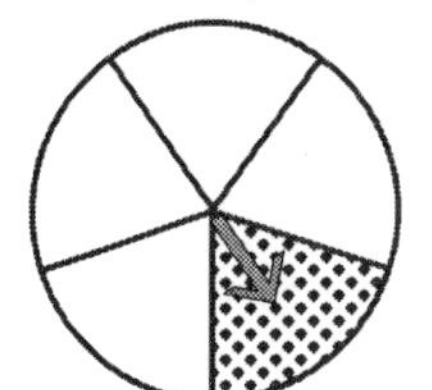

2. Use a centimeter ruler to measure these lines. Write the lengths of the lines to the closest whole number of centimeters.

cm ______
cm ______
cm ______

3. The point (1,3) is marked on the graph. To find (1,3) you start at the bottom left corner of the grid, go → to 1 line, then go ↑ 3 lines. Make dots on the grid for the points below.

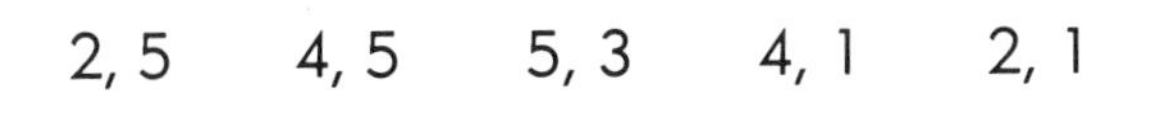

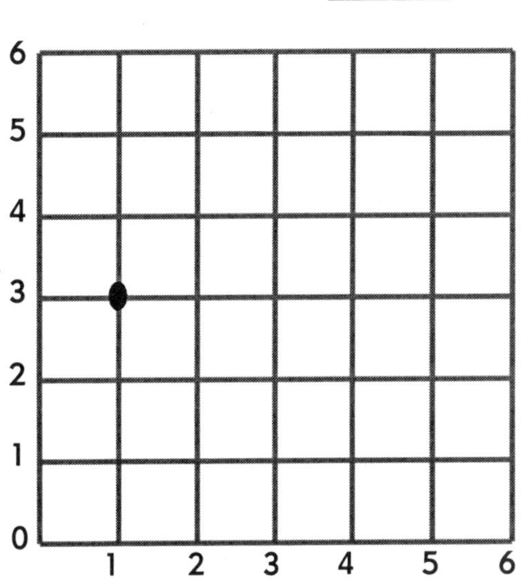

4. Connect the dots, in order, in the grid above. What is the name of the shape you made? ______________________

5. These six small squares can make a rectangle in only two ways.

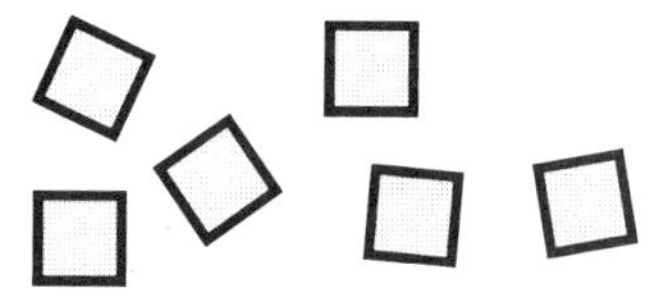

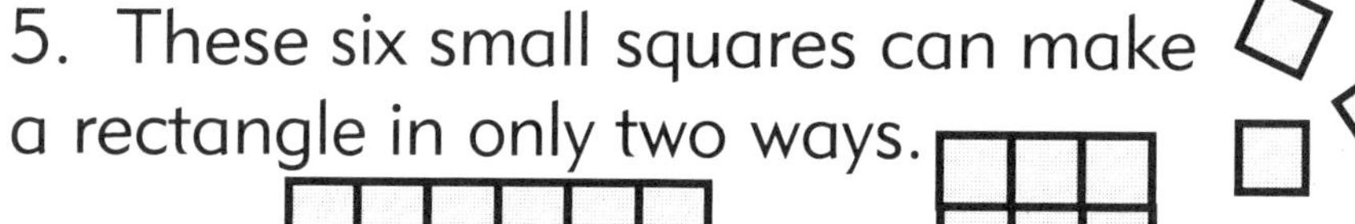

Make as many rectangles as possible using 7, 8, and 12 squares.

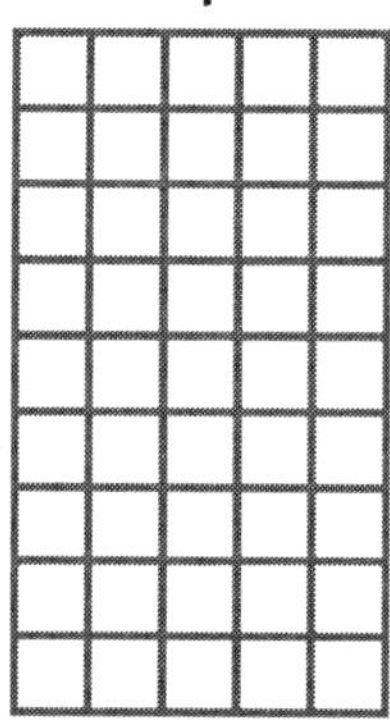
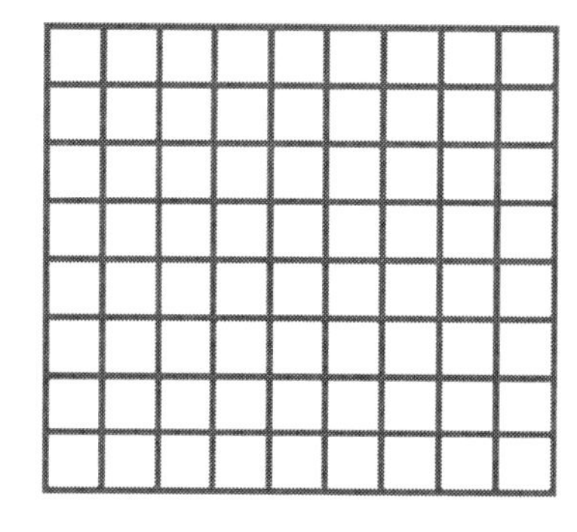
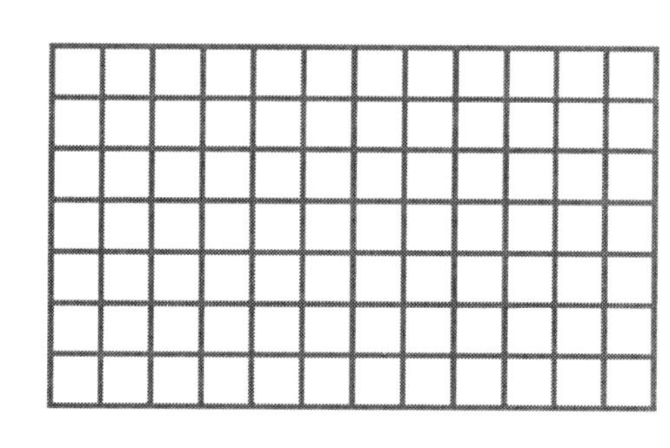

7 squares 8 squares 12 squares

6. Kelly has 3 nickels and 7 pennies. What is the fewest number of coins she could trade for her10 coins and still have the same amount of money? ____________

7. 1/2, 1/3, 1/4 are all fractions that show how much of each circle is filled. Write the correct fraction of the filled space next to each circle.

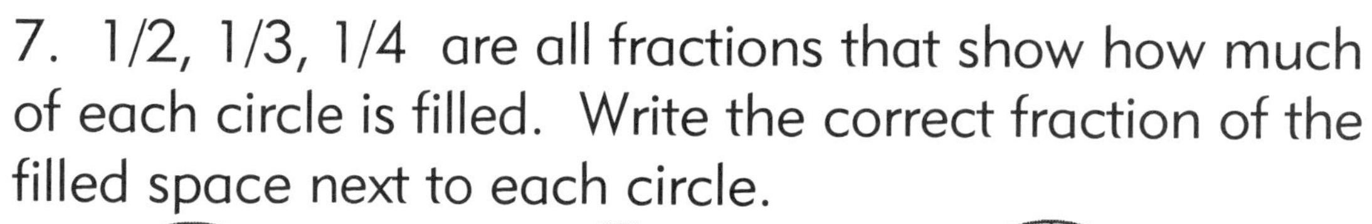

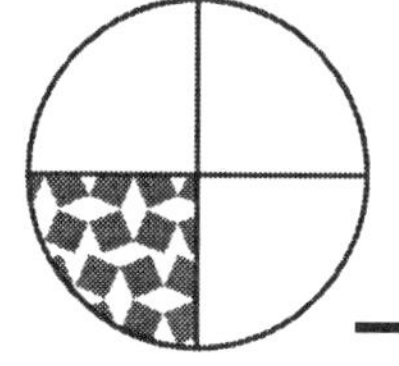 ______ 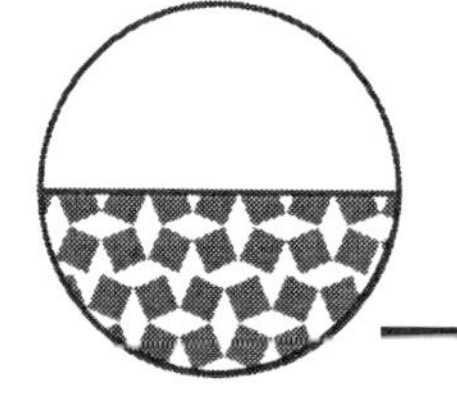 ______ 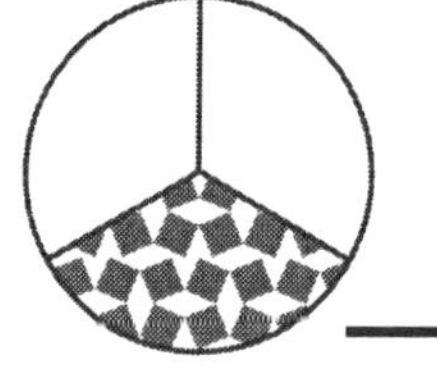 ______

8. Cindy has $2.00 that she can spend at the county fair. Which of these treats can she buy?

lemonade	popcorn	fruit ice	peanuts	balloons
25¢	30¢	$1.50	75¢	10¢

a. 10 ballons & 1 peanuts & 1 popcorn
b. 1 peanuts & 1 fruit ice
c. 1 fruit ice & 1 popcorn
d. 4 popcorn & 4 lemonade

Myself Partners Group

Math Rules!
2: Seventeen

Name ____________________

1. Measure the ribbons.

inches _____ cm _____

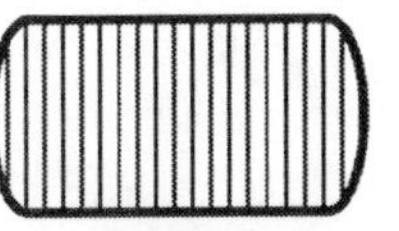

inches _____ cm _____

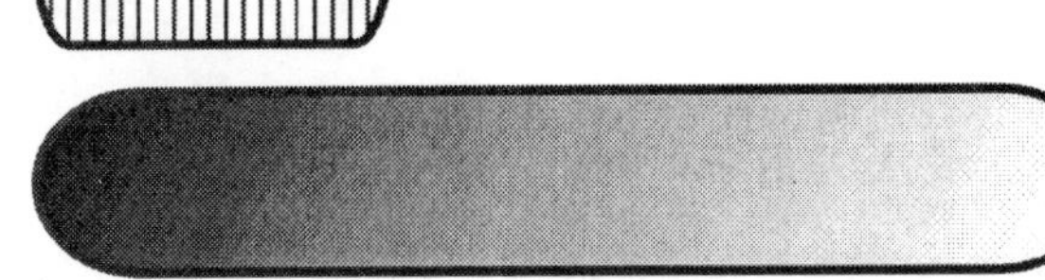

inches _____ cm _____

2. Read the following numbers. Choose only the even numbers and add them together. What is the sum of those even numbers?

twenty-four	thirteen
fifty-seven	forty-four
twenty-six	one-hundred
eleven	sixty-five

3. How many different 2-digit numbers can be made using the digits in the square. A digit may be used only once in a number.

3 7
5 9

________ numbers

4. Arnie has 45¢ in dimes and nickels. He has the same number of dimes as he has nickels. How many of each coin does he have?

__________ dimes and __________ nickels

2: Seventeen

5. One pint  fills 

Circle the number of cups that can be filled with 1/2 pint.

6. This is Brittany's homework page of addition problems. Put a U by the incorrect sums.

1.	2.	3.	4.	5.
24	19	19	28	44
+18	+ 26	+21	+13	+36
42	46	40	40	80

7. Study the pattern and write the missing letters on the grid. Be sure to copy the position the letter is in the pattern.

A	W	E	S		M	
A	W	E		O		E
∀	W		S			

8. Which of these fractions is the largest?

a. 1/2 b. 1/3 c. 1/4 d. 1/5

Math Rules!
2: Eighteen

Name ____________________

1.

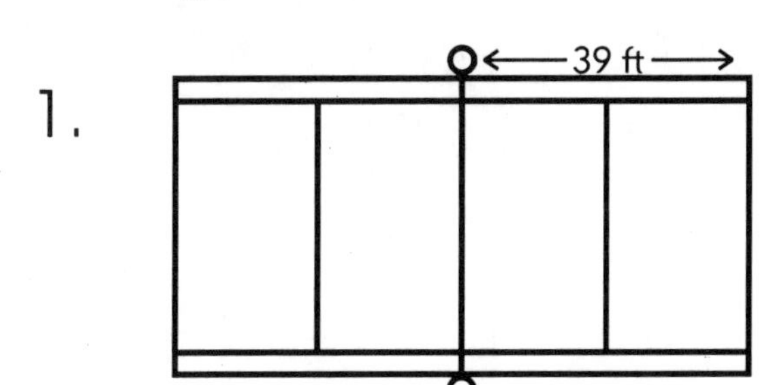

TENNIS COURT

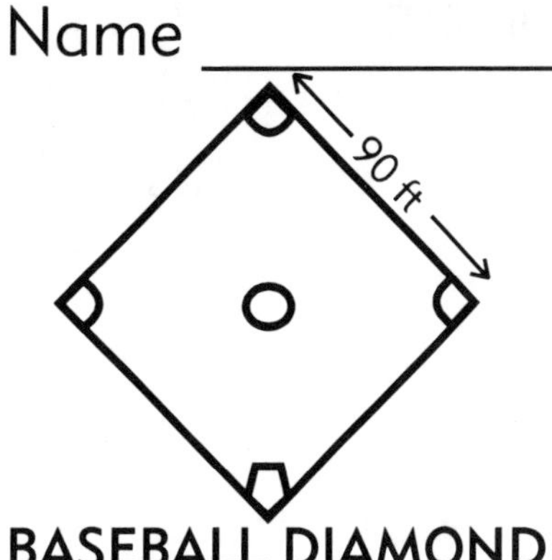

BASEBALL DIAMOND

How far is it from one end of a tennis court to the other end?
What is the perimeter of a baseball diamond?______

2. Use a ruler to draw lines from the point in the center of this triangle, to each of the 3 corners, or angles. The triangle now has how many thirds? ________

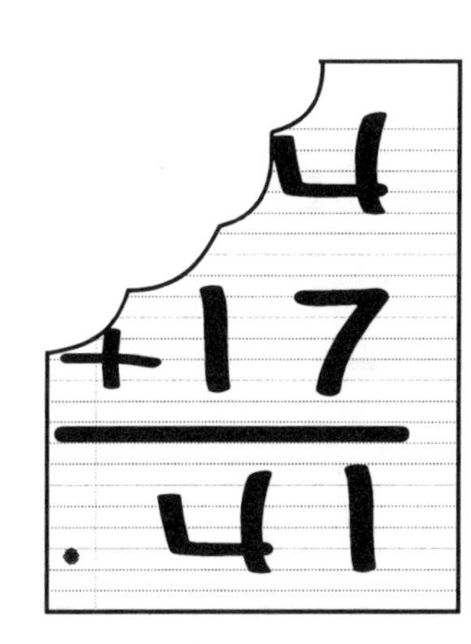

3. "A doggie ate my math paper," Jim told his teacher. Write the missing number in the correct place.

4. It took Ryan's family two days to drive to Disney World. His dad drove 215 miles on Friday and 291 miles on Saturday. About how many miles did the family travel to Disney World? Circle the best estimate.

200 miles   300 miles
500 miles   600 miles

2: Eighteen

5. The Math Rule's mascot can do addition problems quickly in his head. The mascot <u>added the tens place FIRST, and then added the ones place.</u> Practice our mascot's method on these problems.

a. 13 + 52 = _____ d. 53 + 24 = _____ g. 33 + 27 = _____

b. 11 + 18 = _____ e. 41 + 16 = _____ h. 26 + 42 = _____

c. 72 + 17 = _____ f. 46 + 50 = _____ i. 72 + 18 = _____

6. This square was folded in half and folded again in half. When it was opened up, the square had four equal parts.Use dotted lines to show 3 ways that you can show fourths.

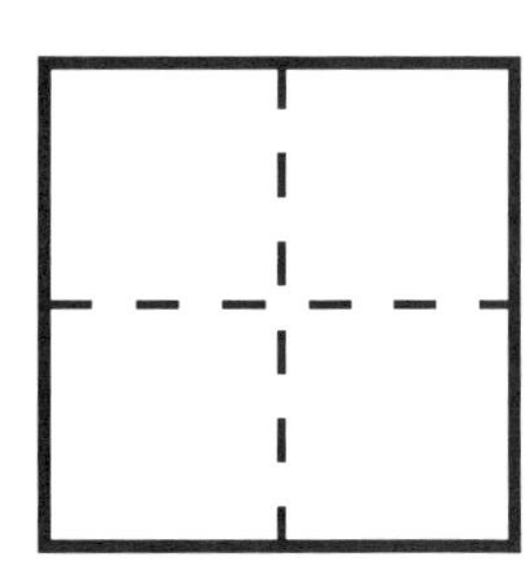

7. Find three consecutive numbers that add up to 45.

consecutive numbers are numbers that follow one another like - 1, 2, 3

The numbers are____,____& ____

8. Circle the best estimate. Tabitha's mom brought a bowl of punch for a class party. About how much punch was in the bowl?

a. 5 meters b. 5 kilograms

c. 5 liters d. 5 bushels

Myself Partners Group

Math Rules!
2: Nineteen

Name ____________________

1. Shapes that make this character:

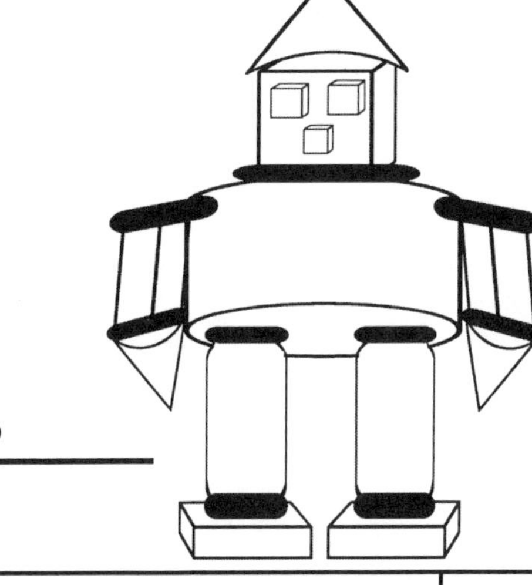

How many cones? ______
How many cylinders? ______
How many cubes? ______
How many rectangular prisms? ______

2. Flip a coin 25 times. Each time you flip the coin make a tally mark by HEADS or TAILS for the side of the coin that lands up. Then add the tally marks to get totals.

		Total
HEADS		
TAILS		

3. Fill in the bar graph to show how many of each heads and tails you flipped in problem 2.

HEADS
TAILS

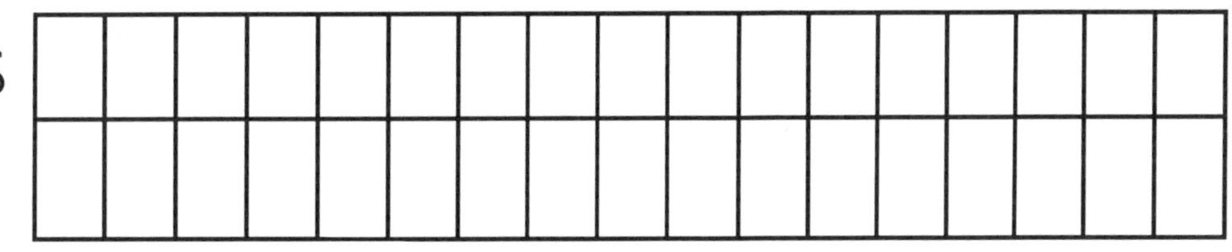

4. About how many pizzas are shown?

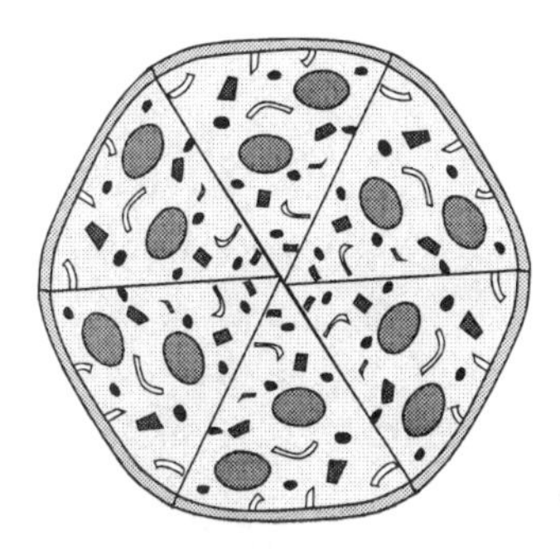
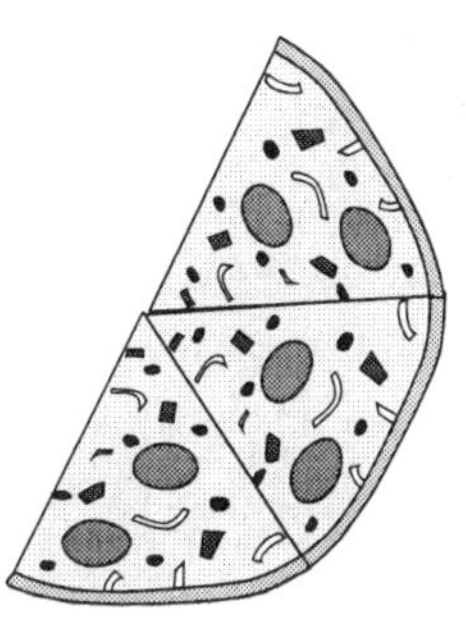
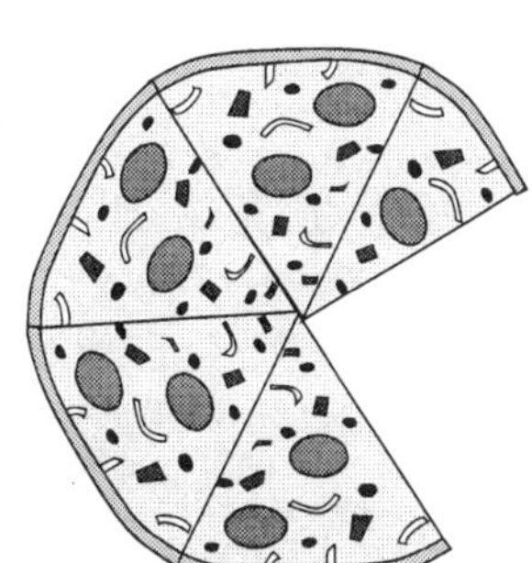

2 pizzas

2 1/2 pizzas

3 pizzas

2: Nineteen

5. How many cubes does it take to build this structure?
_______ cubes

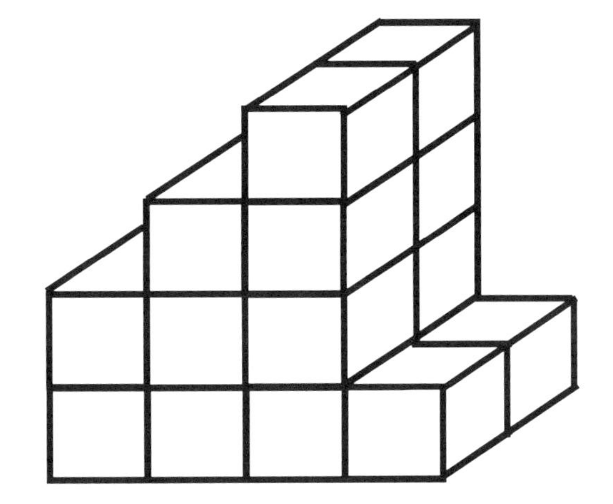

6. To win this tic-tac-toe, find the three numbers that have a sum of 100. Put X 's on those numbers and you win the game.

8	36	43
91	48	22
51	16	67

7. Add.

$$\begin{array}{r} 31522 \\ 44139 \\ +\ 21364 \\ \hline \end{array}$$

8. How much money would you expect to pay for a 5-stick pack of gum at a store where you shop?

a. 10¢ b. $1.75 c. 50¢ d. $1.00

Math Rules!
2: Twenty

Name ____________________

1. Use a calculator to change fractions into decimals.

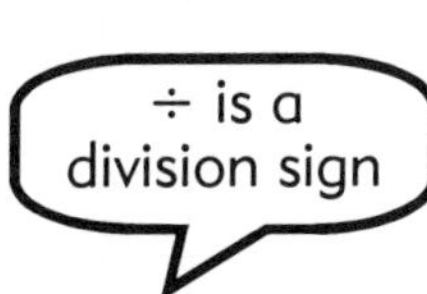

½ is a fraction [ ] is the decimal ¼ is the fraction [ ] is the decimal

2. How many square units make the filled design?

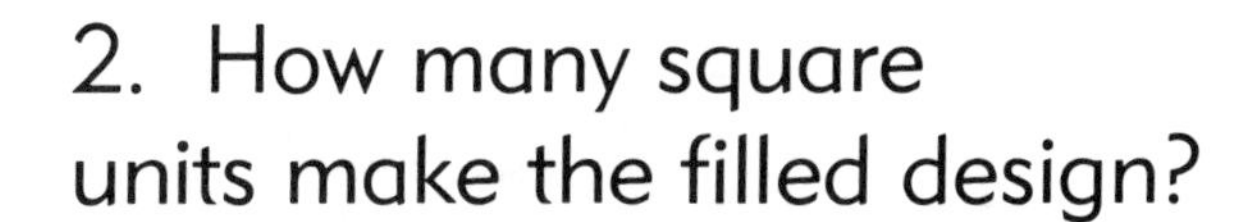

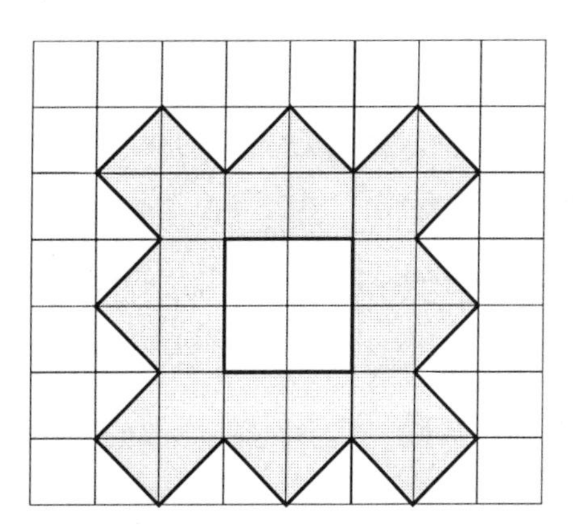

3. There are two numbers whose sum is twenty-one.
One number is 3 more than the other number.
Name the two numbers: ________ & ________

4. Steve's birthday is two weeks from today, on Tuesday the 23rd. What day of the week is today and what is the date? ______________ ______________

2: Twenty

5. These four flat figures can be folded to make solid shapes. Imagine that the design is folded and match them with their name.

a.

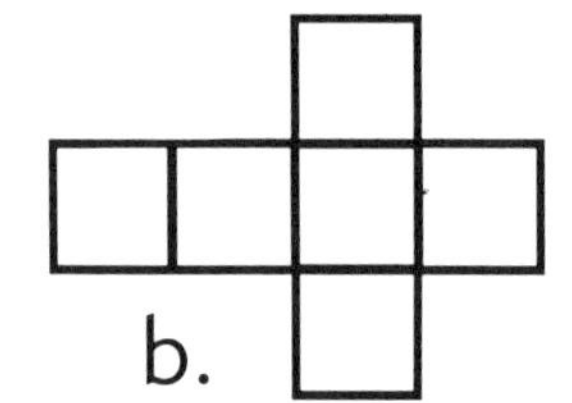

b.

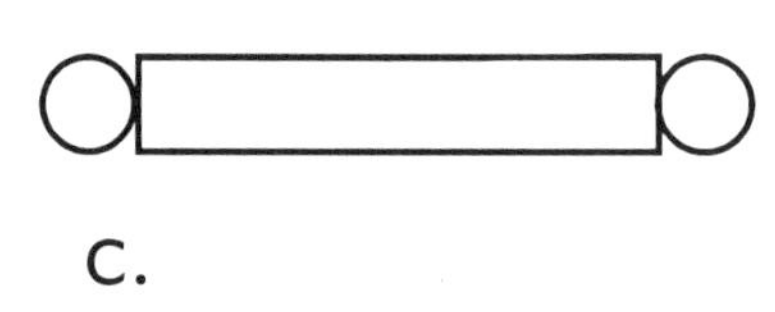

c.

d.

cylinder ______ cone ______
cube ______ rectangular prism ______

6. The temperature at 8:00 in the morning is 58°. If the temperature rises 2 degrees each hour, at what time will the temperature reach 76°? __________

7. Uncle Lou has eighteen boards to use to build a fence between the house and garage. Two boards go between each post. How many posts does he need to nail the boards onto?

_____Posts

8. What solid can not be made from folded flat shapes?

a. sphere b. triangular prism

c. cube d. circle

Math Rules!
2: Twenty-one

Myself Partners Group

Name ____________________

1. Mom is planting tomatoes in the garden. She has 10 plants and she wants to make five rows with four plants in every row. She planted some already. Where will she plant the rest?

2. How many of each coin is in the piggy bank?

=

______ ______ ______ ______

=

______ ______ ______ ______

3. a. Color the path that is 158 footsteps from the barn to the bull RED.

b. How long is the path from the sheep, past the horse, to the silo?

______

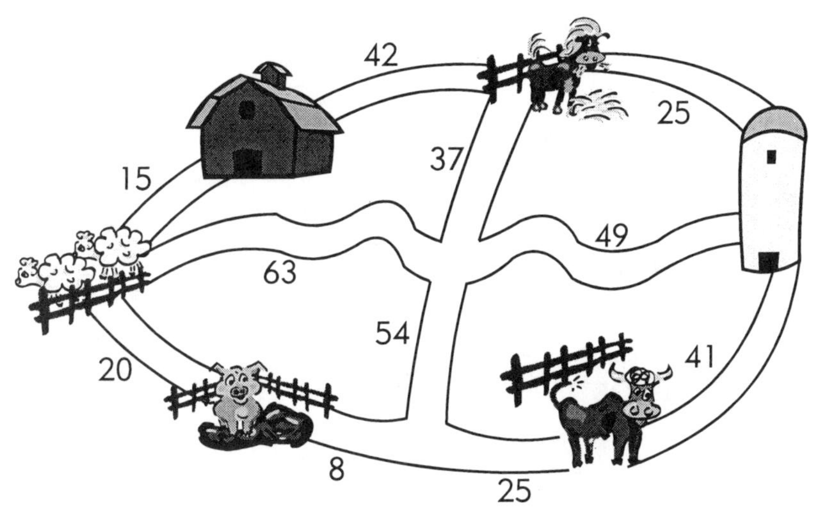

c. Which path is longer from the sheep to the silo? (a) the east path or (b) the southeast path

______

2: Twenty-one

4. Write the number that comes next.

____

5. Match.

TEMPERATURE	CITY	CLOTHING
72	Buffalo, NY	swimsuit
94	Miami, FL	coat & hat
56	Nashville TN	sweater
15	Seattle, WA	T-shirt & skirt

6. This picture is full of parallel lines. Parallel lines are sets of lines that do not cross over each other no matter how long you draw the line. Find 3 sets of parallel lines in this photo and circle them.

BUS SCHEDULE

7. Fill in the table with how long each trip is.

Bus	Bus Leaves	Bus Arrives	Time Trip Takes
Bus 101	4:25 pm	8:00 pm	
Bus 201	11:10am	11:50am	
Bus 301	2:25 pm	4:15 pm	

8. Which number is greatest? Circle it.

a. 5,009 b. 5,950 c. 5,004 d. 5,509

Myself   Partners   Group

Math Rules!
2: Twenty-two

Name ____________________

1.

a. b. c. d. e.

Which two cow pictures are exactly the same? ____ ____

2. Write the name of the object in the column that describes it.

Sphere	Rectangular Prism	Cube

beach ball
aquarium
calculator
Rubic toy
cereal box
toy block
soccer ball
dice
book
orange
globe
balloon
brick

3. What fraction of the students in the class is working on their Math Rules? __________ students

- ➢ There are 12 children in the class.
- ➢ Four students are finished with the assignment.
- ➢ Three students are doing research in the library.
- ➢ The rest of the children are working on their Math Rules!

4. . .542 . . 545 . .547 . . .

Write in the number 544 where it belongs on the number line.

2: Twenty-two

5. Six people want to share 18 muffins equally. How much will each person get?

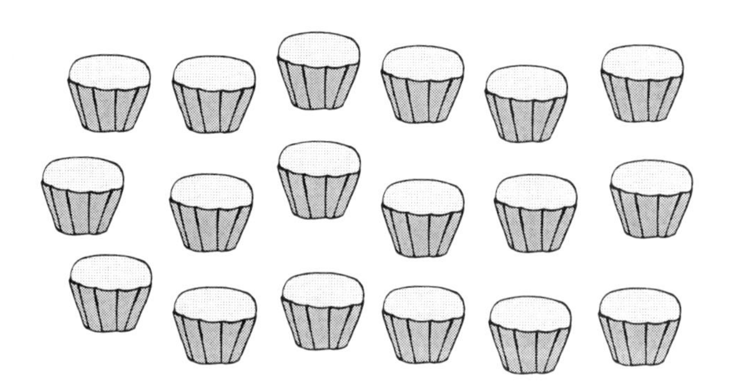

____

6. Put the numbers 0, 10, 20, 30, 40 & 50 in the squares to make the total in each line add up to 100. Use each number only once.

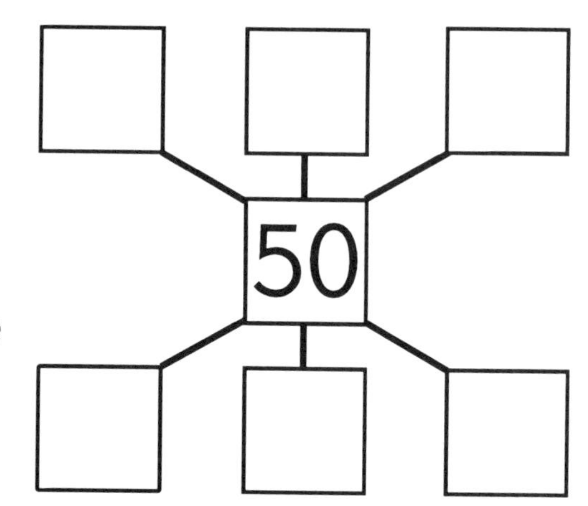

7. Use the code to find the starting number in each row.

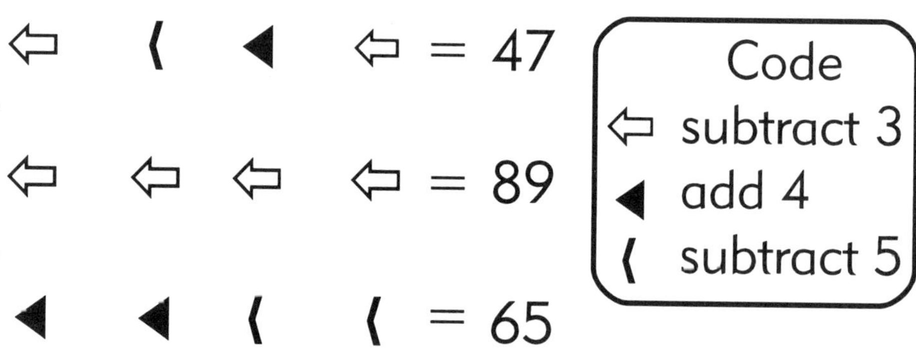

8. Dad used 9 eggs that were left in one egg carton. He also took a full carton from the fridge and added a half-dozen to the recipe. How many eggs are left in the carton?

a. 6 b. 12 c. 15 d. 24

Math Rules!
2: Twenty-three

Name ____________________

1. Circle the best answer.

a.
more than 1 kilogram
less than 1 kilogram

d.
less than 1 kilogram
more than 1 kilogram

b.
about 1 meter
about 1 decimeter

e.
about 1 liter
more than 1 liter

c.
more than 1 pound
less than 1 pound

f.
more than 1 pound
less than 1 pound

2. Use 4, 3, and 2 to write 4 numbers greater than 300.

________ ________ ________ ________

3. Put an X over the die that does not belong.

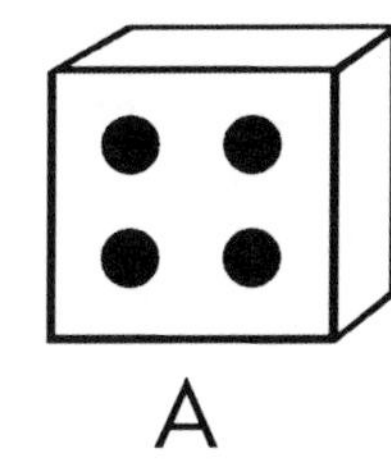
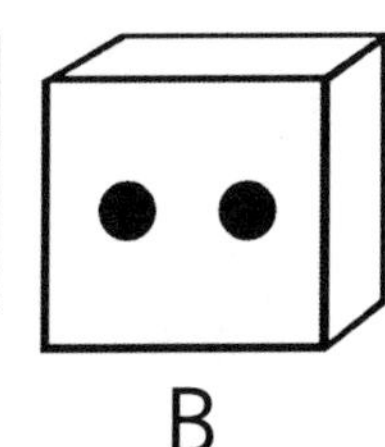
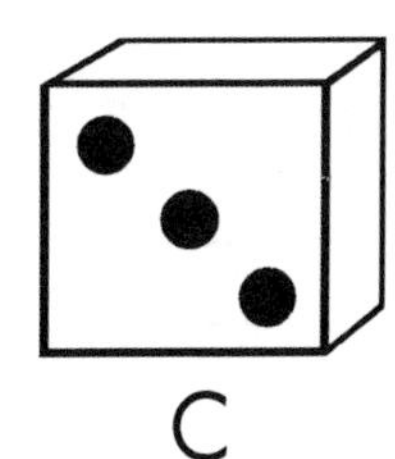

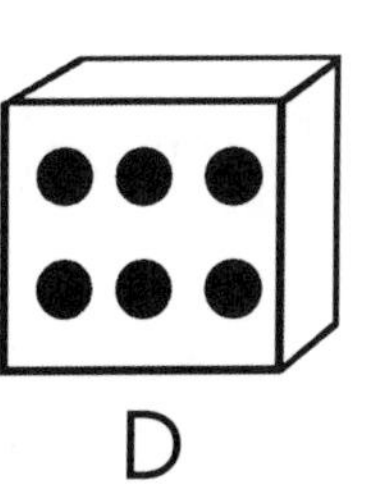

A B C D

4. The school cafeteria has seats for 80 children. One second grade class has 38 children and the other second grade class has 45 children. Can both classes eat in the cafeteria at the same time? __________

2: Twenty-three

5. Put an X over the triangle that does not belong.

a.

b.

c.

d.

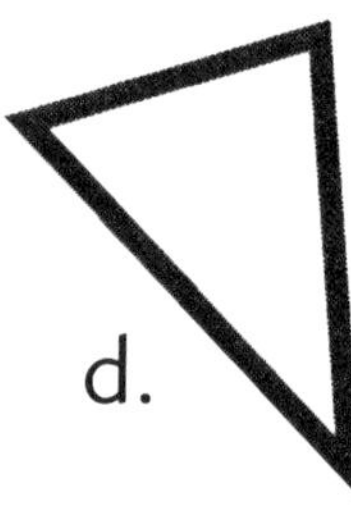

6. Caroline read 36 pages of her 55-page book. Her sister read 73 pages of her 87-page book. Which sister has the most pages left to read? Circle your answer.

Caroline   her sister

7. Seventy-six people traveled on Flight 486. The plane left at 10:37 A.M. Thirty-seven people flew on Flight 210. How many more people are on Flight 486 than on Flight 210? ____________

8. Which thermometer shows a temperature of 32°?

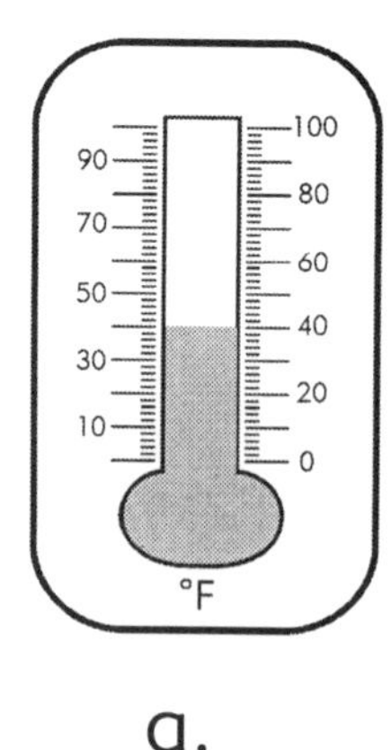

a.

b.

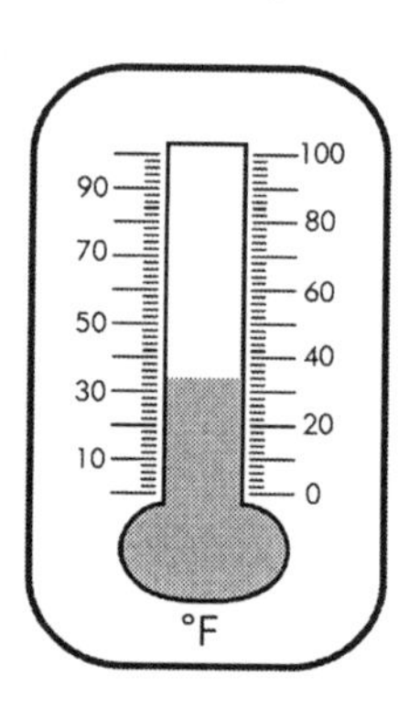

c.

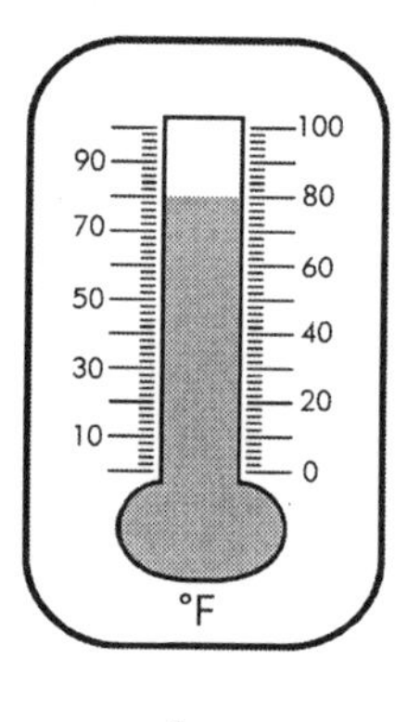

d.

Myself

Partners

Group

Math Rules!
2: Twenty-four

Name ______________________

1. A computer spread sheet can count coins to tell how much money the coins equal. You can calculate the money too. Fill in this spreadsheet.

Student	Number of Quarters	Number of Dimes	Number of Nickles	Total Money
Joe	0	1	2	
Jane	1	1	1	
Jack	1	0	3	
Jerry	0	3	3	

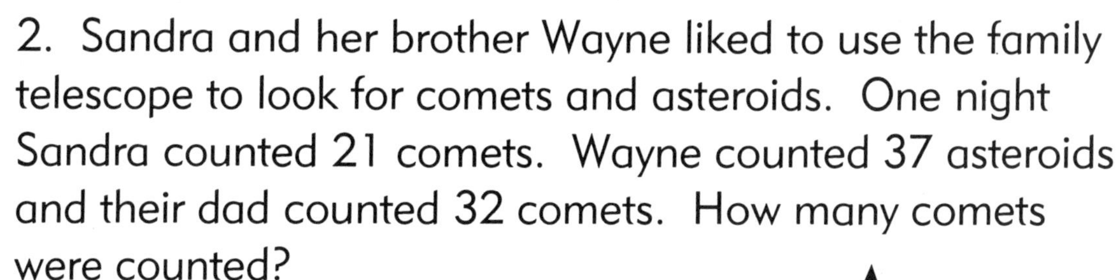

2. Sandra and her brother Wayne liked to use the family telescope to look for comets and asteroids. One night Sandra counted 21 comets. Wayne counted 37 asteroids and their dad counted 32 comets. How many comets were counted? ________

3. Find the Lost Star by going through the correct path. Decide which tens number the number in an area is nearest. Subtract that tens number as you go through the openings until you reach 0. ____________

4. Complete the tables.

	+8
2	
12	
22	
32	

	-7
9	
19	
29	
39	

2: Twenty-four

5. Draw the other half of the design that will match if the paper is folded on the dotted line.

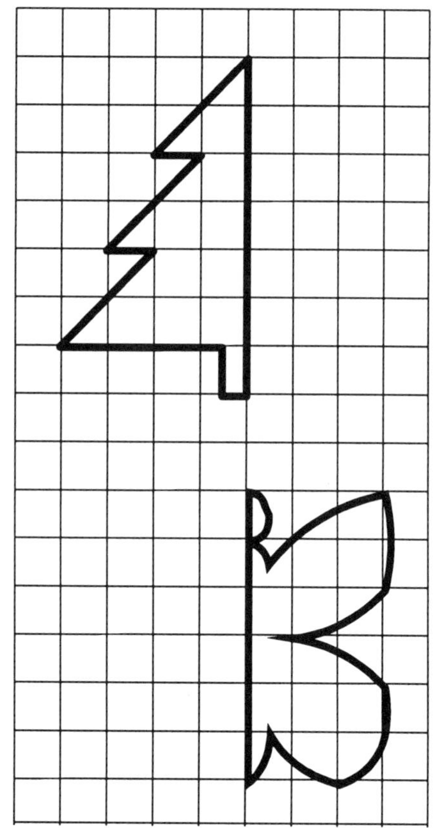

6. Look at these figures called networks. Try to draw a copy of them without lifting your pencil from the paper and without drawing over the same part of line twice. You may cross over a line. Circle the two networks that can be copied this way.

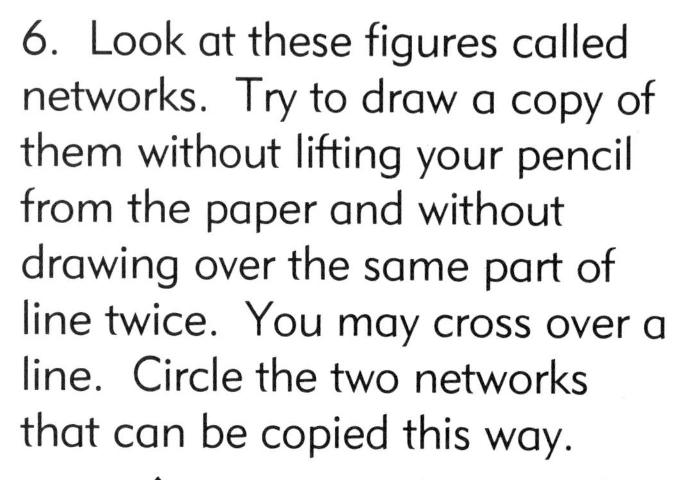

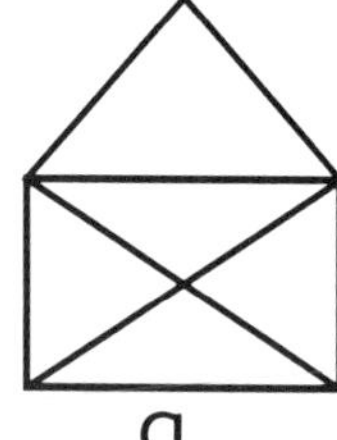
a.

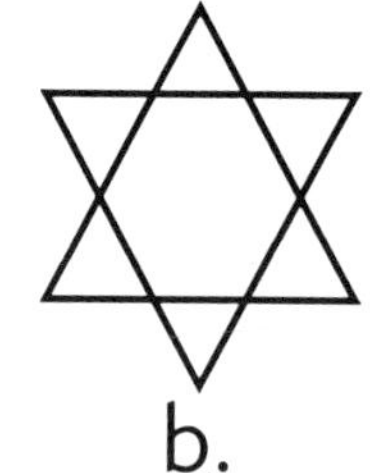
b.

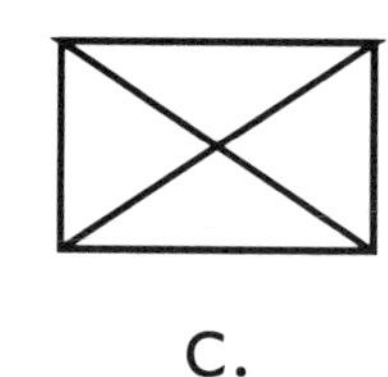
c.

7. The first figure is the same as one of the other three figures, but it is looked at from the other side. Circle it.

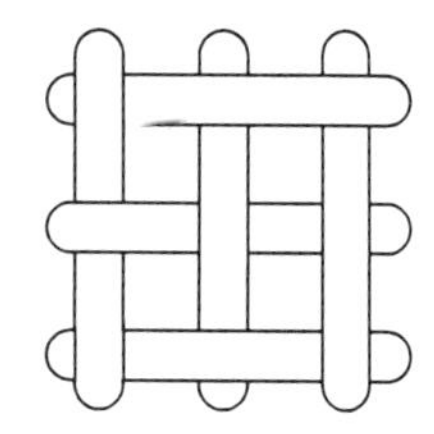

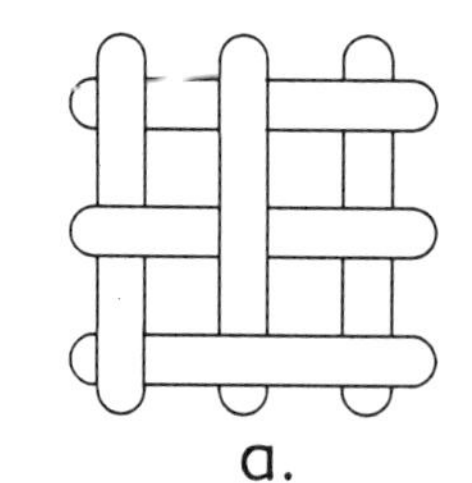
a.

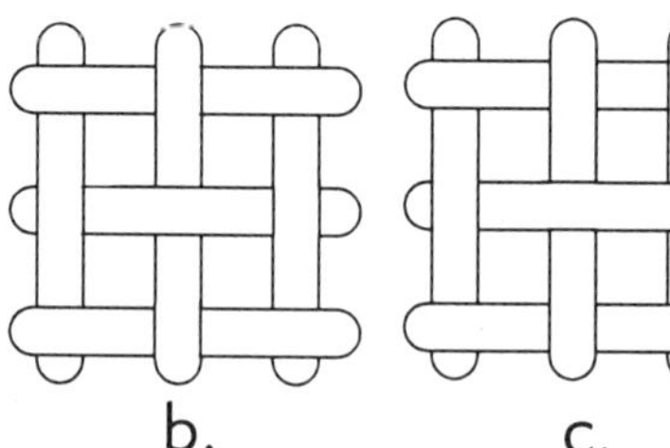
b. c.

8. Take the number 6 and add 5 to it. Double that number, then subtract 4, divide it by 2, and subtract the number you started with. The answer is

a. 6 b. 5 c. 4 d. 3

Myself Partners Group

Math Rules!
2: Twenty-five

Name ______________________

1. Draw three straight lines across the circle. What is the greatest number of areas you can divide the circle into? ____

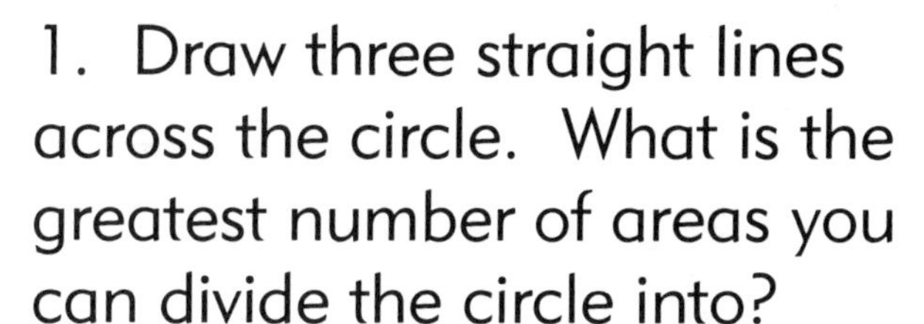

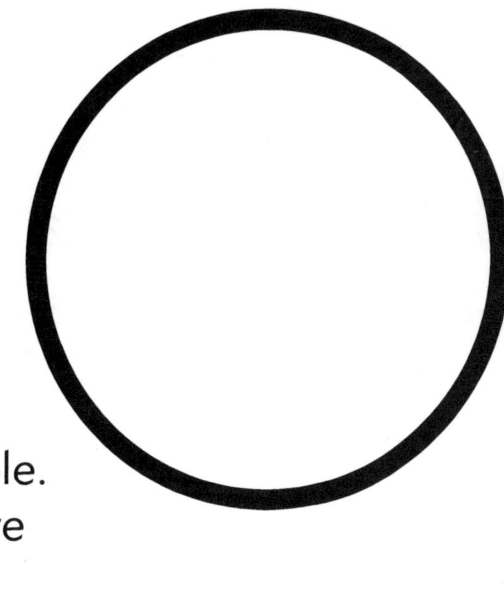

2. Krista earns $ 1.75 each week for setting the dinner table. How much money will she have at the end of three weeks? ____________

3. Tasha worked a math problem on a calculator. She was trying to figure out how old her teacher was. Look at the display on this calculator. Could her answer be correct?

YES NO

4. A cat tried to climb a 10 foot tree. He started to climb and was able to climb 3 feet in one minute. But, the curious cat stopped to look around and fell back one foot. He did this same thing each minute. How many minutes did it take for the cat to reach the top of the tree? ____________

2: Twenty-five

5. An odd number comes before 50 and after 50. If you add these two odd numbers together, what is the total? _______

6. Match the fraction with the shaded part of the design.

1 ______
1/2 ______
1/3 ______
1/4 ______
1/5 ______
1/6 ______

7. If you counted eleven kids ahead of you in the dismissal line at school and counted twelve kids behind you, how many kids are in the line waiting to go home?__________ kids

8. Circle the list in which 2/3 of the words mean the same thing.

a.	b.	c.	d.
tiny	red	run	over
huge	white	skip	under
small	blue	jump	behind
		purr	

ANSWERS

1: One

1. 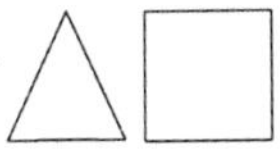
2. All but the bird
3. 7 squares filled in each grid.
4. No
5. 40¢
6. 10

1: Two

1. 3, 4, 9, 10
2. 2, 4, 6, 3, 9
3. Second
4. tenth - J second - B eighth - H sixth - F
5. Sarah
6. 6 + 5= 11

1: Three

1. 12¢, 8¢, 4¢
2. left bucket
3. nine dots
4. 3rd
5. 42 stars
6. 3

1: Four

1. answers vary
2. 8
3. 3
4. Possible configuration

    1
    6 4
    3 2 5

5. 5 each hand
6. 5+2=7 1+4=5 4+3=7 2+2=4

1: Five

1. farthest left and 3rd from left
2. 8¢
3. Lions - 10; Tigers - 9
4. dots connect to make bird
5. 
6. 20, 25, 23, 20, 28

1: Six

1. 8 clips
2. 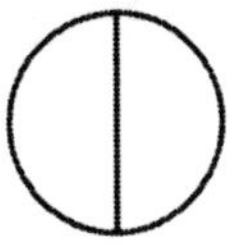 Possible answer
3. 16
4. ✪ 24 ❋ 23 ❄ 19 ✳ 15
5. 10 °F 40 °F
6. Check student answers

1: Seven

1.

    1 2 3
    4
    5 6 7

2. 253
3. butterfly, flower, ladybug
4. greater, greater, less, greater, less, greater, less
5. 4 hours
6. 5 squirrels

1: Eight

1. 245-3746 647-9338 858-6284
2. 7
3. Triangle - 7 dots square - 5 dots hexagon - 6 dots
4. 4848 9348 6783 7557 6666
5. boy - 13 cm arm - 5 cm shoe - 2 cm
6. 9, 8, 4, 3 (outside ring); 4, 7, 9, 2 (inside ring)

1: Nine

1. 4-1 1-4 3-2 2-3 5-0 0-5
2. 
3. 12 cm
4. 1st floor - Miles; 2nd floor - Aaron; 3rd floor Grandma; 4th floor - Bryan; 5th floor - Ann
5. B
6. 5th row - 6 cards; in all - 40 cards

1: Ten

1. 8 cats 6 dogs 3 birds 3 turtles
2. a. 1 b. -1 c. 2 d. 3 e. -2 f. -4 g. 3
3. Check student answer

4. 9 angles
5. Ted
6. 1st box 2, 9, 4 2nd box 18, 28, 21, 25
 3rd box 49, 63, 99

1: Eleven
1. subtract, zero, fifth, sum, time
2. 6-9-4 7-9-3 8-9-2
3. 8 triangles
4. 36, 33, 34, 31 Cierra
5. flower and spider
6. 32 squares

1: Twelve
1. 8 meters
2. ~~26~~/24 ~~58~~/54 ~~68~~/64 ~~81~~/84
3. 2 cookies
4. ~~31~~/30 ~~27~~/26 ~~21~~/20 ~~15~~/16 ~~7~~/8
5. answers vary
6. 7735 SELL

1: Thirteen
1. 3, 4, 5, 6
2. answers vary
3. meters
4. 23 fruits
5. 13 cm long; check student answer
6. 6

1: Fourteen
1. 2nd box from right
2. 16-17; 25-26; 31-32
3. 20, 25, 30, 35
4. 15 pennies; 5 in a row; 3 in a column
5. b
6. 3 apples; 2¢ left over

1: Fifteen
1. 4 times
2. 4-9-2 3-5-7 8-1-6
3. dotted
4. 12 donuts
5. 60°; 40°; 72°
6. cool; cold; warm

1: Sixteen
1. any amount over 5
2. a. 12 b. 14 c. 14 d. 22
3. - 8:00 - 3:30
4. 20 spaces
5. 7 tens, 7 ones
6. 3 dimes crossed; 3 pennies crossed

1: Seventeen
1. 10 animals free
2. 12 cookies
3. B and C
4. about 7-8 dimes
5. 8
6. 12 inside 24 outside, 6 on
 17 inside 8 outside 8 on

1: Eighteen
1. less than 2 decimeters; less than 2 decimeters
2. 55¢
3. 18, 20, 15, 18
4. 14
5. Wednesday
6. At least 10 ways

1: Nineteen
1. .04, 57¢; .06, 91¢; .00, 10¢;
 .02, 27¢; .05, 84¢
2. answers vary
3. answers vary
4. $18; 75¢; $24
5. a. 40 b. 60 c. 50 d. 40
6. 35 flies

1: Twenty
1. 8, 21, 15, 32
2. 72, 53, 44, 95
3. 9 cards
4. 12:30
5. a. 6 b. Motor c. 5
6. TERRIFIC

1: Twenty-one

1.

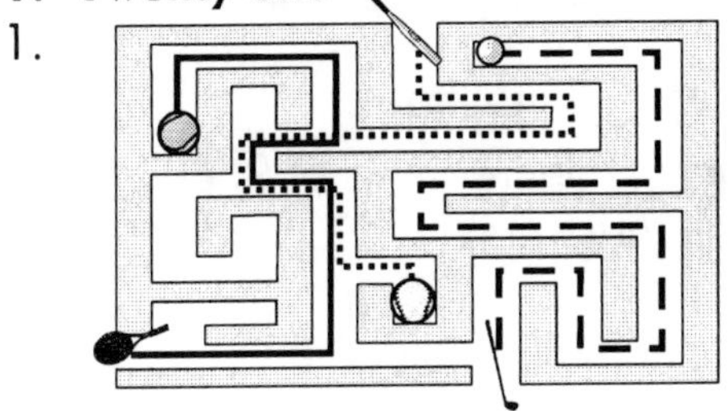

2. 90¢
3. 20 pretzels
4. Strawberry, apple, bananas, watermelon, potatoes
5. 8 tires
6. 6, 12, 8

1: Twenty-two

1. 7 pounds
2. 42

2	4	8	33	62	41	27	29
3	6	9	12	19	13	15	11
51	8	10	15	37	24	22	20
13	15	12	18	21	24	17	35
5	24	14	16	18	27	40	38
45	57	24	36	20	30	33	39
30	28	26	24	22	48	36	56
32	34	36	38	40	42	39	0

3. 5 inches
4. A. 30 B. 40 C. 60 D. 50 E. 70 F. 60 G. 20 H. 80
5. scales - meat, TV, nut; ruler - rope; measuring cup - fish bowl, bucket
6. 12

1: Twenty-three

1. 3rd from right
2. 10¢ 5¢ 5¢
3. 73, 63, 53, 43, 33, 23, 13, 3
4. about 3 feet high
5. =
6. 6 tens 4 ones 2 hundreds

1: Twenty-four

1. 13-7=6; 8-4=4; 5-3=2
2. A. 3 o'clock B. 6 o'clock C. 1 o'clock D. 7 o'clock
3. 1/3 1/2 1/4

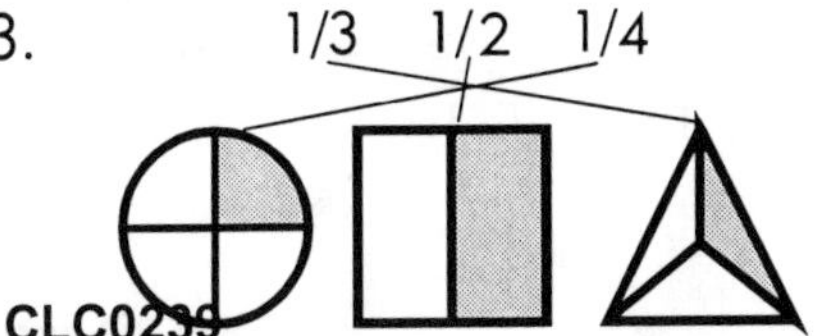

4. 7, 12, 9, 13, 18, 16, 13, 11, 15
5.

4	5			5	6
2	6			2	7
		8	4		
		3	1		
6	9			1	7
9	9			2	1

6. 14

1: Twenty-five

1. dotted shirt, white shorts; stripped shirt, white shorts; stripped shirt, solid shorts; dotted shirt, solid shorts
2. a. 14 b. Wednesday c. 31
3. b
4. 14
5. 3 sets of 6 balloons
6. black

2: One

1.

2

3 4

1 5 0

Possible configuration

2. 9 cents
3. 12 OR 20 legs
4. 45
5. 6 ; 5; - ; 6
6. 3, 6, 5
7.
8. c. 11

2: Two

1. all, some, some, none
2. Bac - 5 Tac - 3 Mac - 7 Clack - 2
3. a. 3-4-5 b. 11-16-17
   c. 23-19-15 d. 3-9-13-19
4. 15 - pentagon
5. top clockwise: 7, 6, 5, 4, 3, 2, 1, 0
6.

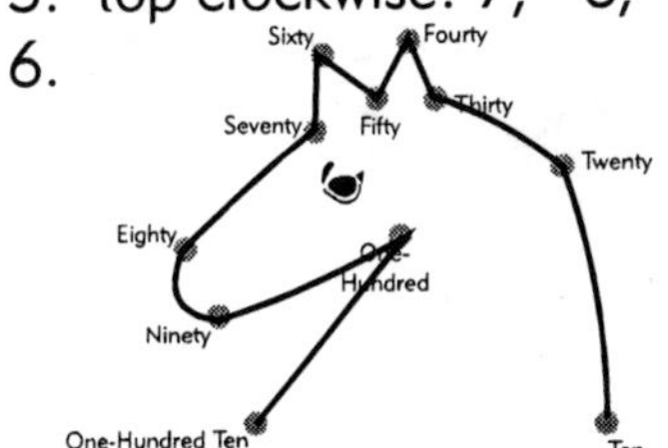

7. 6 paths
8. b. 4

2: Three

1. a. 11 b. None c. Thursday d. ◆◗◗
2. a. 43 b. 38 c. 42
3. 13 - odd; 14 - even; 18 - even;
   19 - odd; 25 - odd
4. X, IV, varies, IX
5. 38 48 64 77
6. 77 64 48 38
7. Check to see that touching states are different colors
8. c. 56

2: Four

1. lollypop; got a nickel and penny back
2. □
3. 42
4. match coins to each choice of ride
5. 42 apples
6. MATH IS POWER
7. 49, 50
8. a. 6

2: Five

1. 
2. 562, 1096, $10.29
3. 3 boxes
4. 12
5. $3
6. 6, 10, 6
7. <, <, <
8. b. 2 bags of 20

2: Six

1. 14
2. 4+3+2=9; 6+4-2=8; 6+4-3=7; 6-4-2=0
3.

```
     1
   4 3 2    Possible answer
     5
```

4. Check that drawn line is perpendicular to line.
5. .10
6. 55, 64, 73, 82, 91
7. all sums equal 10, increments of 9, odd/even repeated (possible answers)
8. b. Friday

2: Seven

1. dots are 4 X 4
2. 16th (Saturday)
3. 4:10; 5:25
4. 74 > 57
5. Check that graph reflects tally marks.
6. About 13 cm
7. 18
8. a. N

2: Eight

1. ☹
2. a. 1 cm b. 4 cm c. about 7 cm d. 2 cm
3. Delete 9 from C; add it to A; all = 15
4. 33; 24; 40
5. 5, 5; 2,1; 4,6; 4,2
6. 427-83=344
7. 34 circles
8. d. 9

2: Nine

1. Mar.; third; May; fifth; September; Sept.; Nov.; eleventh
2. 6
3. yellow fish surrounded by blue
4. 8
5. 4
6. most - purse c; least - purse b
7. $1.92
8. d. $

2: Ten

1. 21 cubes
2. 4:15
3. 20 degrees, 7 degrees, -10 degrees
4. B. less than 5 pounds
5. Check that shape and size are the same.
6. a. 25 b. 29 c. 14 d. 13

7. 8
8. d. 188 cm

2: Eleven
1. 3 cylinders 1cone 5 spheres 4 cubes
2. N, M, J, K, D, L, B
3. U BELONG
4. 8 milk cartons
5. 13 triangles
6. 1+2-3=0; 6-5+4=5; 9-8+7=8
7. 8 cups in all; 4 pints in all; 2 quarts in all
8. A. 202 B. 11 C. 335 D. 65 c. SEE

2: Twelve
1. < 1 pound, = 1 pound , > 1 pound,
   < 1 pound, > 1 pound, = 1 pound
2. top clockwise: 12, 8, 5, 7, 11, 6, 10, 9
3. Chelsea had 1 more marble than Zach
4. 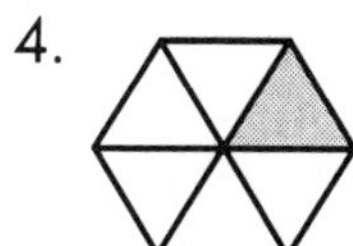
5. h, b, d, c, f, e, g, a
6. striped 16-4=12; solid 8+4=12
7. 7, 9, 4
8. c. liters

2: Thirteen
1. Answers vary with reasons
2. 35006; goose
3. 15,0,0; 5,2,0; 5,0,1; 0,3,0; 0,1,1; 1 0,1,0
4. less than a liter
5. a. cat b. horse c. cat, dog, pig, goat
6. horse, cow, pig, dog, goat, cat
7. 35, 38, 41, 44, 47, 50;
   82, 86, 90, 94, 98, 102
8. b.

2: Fourteen
1. false, true, true
2. a. 8 blocks b. tree; about 4 blocks c. 4 blocks
3. 30
4. 93¢
5. 2 for 12-6¢, 2 for 8-4¢, 2 for 18-9¢,
   2 for 10-5¢, 2 for 16-8¢, 2 for 14-7¢
6. 8 "shoes" long

7. 70°F, 80°F, 50°F, 10°F
8. d. 210 minutes

2: Fifteen
1. 20 cm
2. b. and d.
3. 4th red, 12th green, 19th brown
4. a. 567 b. 823 c. 134
5. 9
6. 6, 5
7. 1. M 2. I 3. L 4. B 5. D 6. CA
8. b. 5

2: Sixteen
1. 1/3, 1/5
2. 2 cm, 8 cm, 11 cm
3. 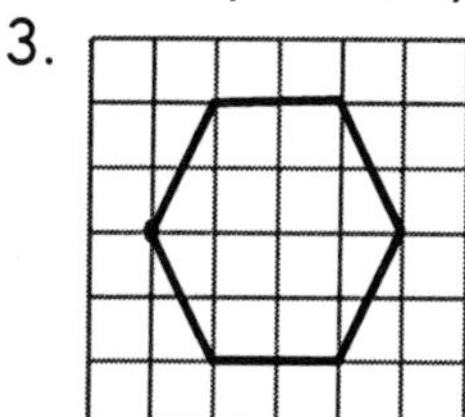
4. hexagon
5. Check student answers
6. 2 dimes & 2 pennies
7. ¼, ½, 1/3
8. c. 1 fruit ice & popcorn

2: Seventeen
1. 2 inches - 5cm 1 inch - 2.5cm
   3 inches - 7.5cm
2. 194
3. 12 numbers
4. 3 of each coin
5. 1 cup
6. 2, 4
7.

A	W	E	S	O	M	E
A	W	E	S	O	M	E
A	W	E	S	O	M	E

8. A. 1/2

2: Eighteen
1. a. 78 feet b. 360 feet
2. 3

3. 2
4. 500
5. a. 65 b. 29 c. 89 d. 77 e. 57 f. 96
   g. 60 h. 68 i. 90
6. Possible configuration
7. 14, 15, 16
8. c. 5 liters

2: Nineteen
1. 3 cones; 12 cylinders; 4cubes;
   6 rectangular prisms
2. answers vary
3. graph must correspond to tally marks in problem 2.
4. 2 1/2 pizzas
5. 20 cubes
6. 36-48-16
7. 97025
8. c. 50 cents

2: Twenty
1. 0.5; .25
2. 22 ☐s
3. 9; 12
4. Tuesday, 9th
5. A. Cone B. Cube C. Cylinder
   D. Rectangular prism
6. 5:00 P.M.
7. 10 posts
8. a sphere

2: Twenty-one
1.

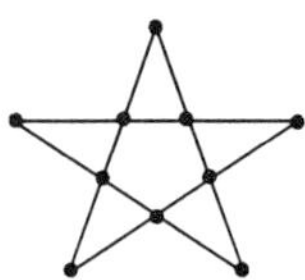

2. 25¢/2, 10¢/1, 5¢/1, 1¢/0;
   25¢/4, 10¢/1, 5¢/1, 1¢/3
3. b.125 steps; c. east path
4. 16
5. 72° - Nashville - T-shirt & skirt
   94° - Miami - swim- suit
   56° - Seattle - sweater
   15° - Buffalo - coat & hat

6. Many choices available.
7. 3 hours, 35 minutes; 40 minutes;
   1 hour, 50 minutes
8. b. 5,950

2: Twenty-two
1. c & e
2. sphere - beach ball, soccer ball, balloon, orange, globe; rectangular prism - aquarium, brick, cereal box, calculator, book; cube - dice, Rubic toy, toy block
3. 5/12
4. Check student answers
5. 3 muffins
6.
   20 - 10 - 0
   50
   50 - 40 - 30
7. 54, 101, 67
8. a. 6

2: Twenty-three
1. A. less than a kilogram B. about 1 decimeter
   C. more than a pound D. Less than a kilogram
   E. More than a liter F. Less than a poung
2. 342, 432, 324, 423
3. C
4. No
5. C
6. Caroline
7. 39 people
8. c. 32°F

2: Twenty-four
1. Joe-20¢ Jane-40¢ Jack-40¢ Jerry-45¢
2. 53 comets
3.

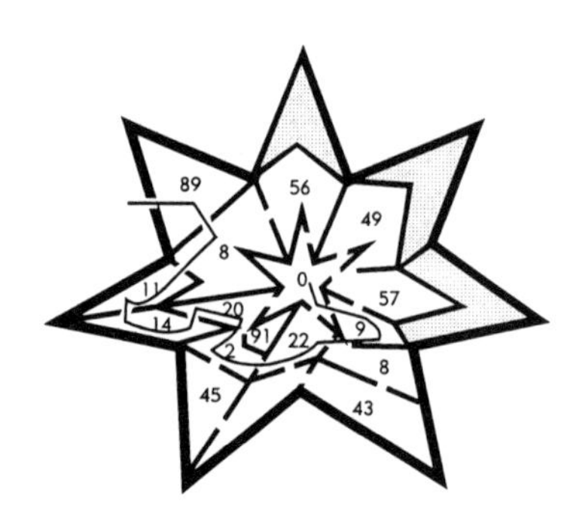

4. +8 = 10, 20, 30, 40; -7 =2,12,22,32

5. 

6. a & b
7. c.
8. d. 3

2: Twenty-five

1. 7

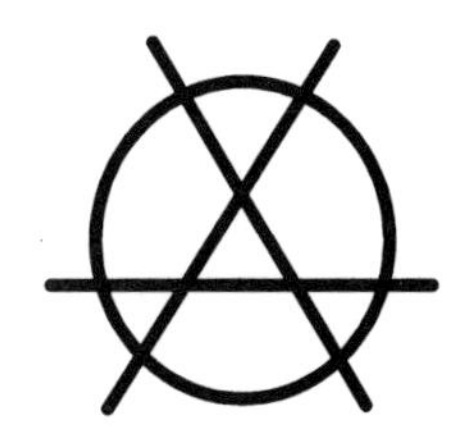

2. $5.25
3. No
4. 5 minutes
5. 100
6. e, d, b, c, f, a
7. 24 kids
8. a. tiny, huge, small

NOTES

NOTES